ADVANCE PRAISE FOR

# A BOOK *of* ECOLOGICAL VIRTUES

"Here is a compendium of capacious hearts, a bundle of essays written by an array of luminous animals, each articulating the deep beauty of things while grappling with their own responsibility for the rapid and accelerating diminishment of that beauty. This is a book filled with *soul*—the ability to feel with and for the world in all its particularity and multiform strangeness. It bespeaks a new kind of scholarship, at once rigorous and richly empathic, participatory, and vulnerable."

— David Abram, author of *Becoming Animal* and *The Spell of the Sensuous*

"[For] those of us who are thinking about, and educating for, deep cultural change,...for those of us that care about what it means to be good, in the deepest sense; to participants in an earth-system that is failing at human hands, in the context of recognizing the repercussions of the Anthropocene."

— Laura Sewall, author of *Sight and Sensibility: The Ecopsychology of Perception*

"*A Book of Ecological Virtues* is a book for our times—to help us recognize that the extinction crisis we face is a consequence of seeing ourselves as separate from nature and other beings. [It] is a remembrance that we are of the earth, we are animals, we are biological beings, nourished by other beings. Caring for others is our ecological duty and the path to our own well-being."

— Vandana Shiva, author of *One Earth, One Humanity vs The 1%*

"A tour de force [that] addresses morality and sustainable futures for humanity. Diverse scholarly perspectives, using a variety of representations, raise issues concerning transformation of human lifestyles and interactions with living and nonliving constituents of fragile ecosystems. Educating a citizenry to act with heightened awareness, compassion, and love are among issues addressed throughout the book. Also, engaging critical dialogues around issues such as death and extinction are raised in a context of how to act courageously, with overt consciousness and an ethic of giving back. The book is captivating, thought-provoking, and forward looking—an amazing graduate text, and an interesting read for a diverse variety of educators, researchers, and policymakers."

— Kenneth Tobin, editor of *Mindfulness in Education*

# A BOOK *of* ECOLOGICAL VIRTUES

*Living Well in the Anthropocene*

*Edited by* HEESOON BAI, DAVID CHANG,
*and* CHARLES SCOTT

University of Regina Press

Cover art: Photo by Dave Hoefler on Unsplash
Cover design: Duncan Campbell, University of Regina Press
Interior layout design: John van der Woude, JVDW Designs
Copy editor: Ryan Perks
Proofreader: Kirsten Craven
Indexer: Jason Begy Indexing

**Library and Archives Canada Cataloguing in Publication**

Title: A book of ecological virtues : living well in the anthropocene / edited by
    Heesoon Bai, David Chang, and Charles Scott.
Names: Bai, Heesoon, editor. | Chang, David, 1978- editor. | Scott, Charles
    (Charles Farquhar), editor.
Description: Includes bibliographical references and index.
Identifiers: Canadiana (print) 20200245805 | Canadiana (ebook) 20200245988 |
    ISBN 9780889777569 (softcover) | ISBN 9780889777620 (hardcover) |
    ISBN 9780889777583 (PDF) | ISBN 9780889777606 (HTML)
Subjects: LCSH: Environmental ethics. | LCSH: Ecology—Philosophy.
Classification: LCC GE42 .B66 2020 | DDC 179/.1—dc23

University of Regina Press, University of Regina
Regina, Saskatchewan, Canada, S4S 0A2
TEL: (306) 585-4758 FAX: (306) 585-4699
WEB: www.uofrpress.ca

We acknowledge the support of the Canada Council for the Arts for our
publishing program. We acknowledge the financial support of the Government
of Canada. / Nous reconnaissons l'appui financier du gouvernement du
Canada. This publication was made possible with support from Creative
Saskatchewan's Book Publishing Production Grant Program.

*We wish to dedicate this volume to the life and work of our colleague Carl Leggo. His presence, pedagogy, writing, and boundless generosity of heart touched the lives of thousands. We are mindful that his chapter here, beautifully co-written with Margaret McKeon, is among the last of his published works.*

# Contents

# Acknowledgements

We editors would first like to thank all our authors, whose exceptional work was matched by their generosity, kindness, and patience as we laboured through initial revisions, peer review, and various additional revisions that included cuts to text and the back-and-forth of finalizing changes to *Chicago* referencing for those who were used to APA. We are very grateful for your work and your magnanimous spirits!

Thank you to Karen Clark for your wonderful support as editor. We appreciate your patience in dealing with our many queries and countless emails. Thank you also to Kelly Laycock for her expert and patient work in managing the project. We also wish to express our appreciation to Ryan Perks and others in the team of copy editors and marketing associates. Our authors repeatedly expressed their appreciation of the care and attentiveness shown by the entire team.

Thank you to Cameron Duder for your expert copy-editing and your work on all the references. We appreciate your unstinting labour under pressing deadlines.

We would also like to recognize the support of our families, who tolerated our absences as we toiled over this project. Your patience has made this book possible; its fruits issue from your largess.

Finally, we bow in gratitude to the earth itself, whose bounty sustains our labour, whose beauty inspires our virtue.

# The Call for an Ecological Virtue Ethics

HEESOON BAI, DAVID CHANG,
& CHARLES SCOTT

D ESPITE OUR BRIEF TENURE ON PLANET EARTH, *HOMO SAPI-ens* have yet to become *sapiential.* In the collective pursuit of *progress,* humanity has reached the point of fouling its own nest and imperiling the basis of life on earth. We have now entered the newly designated and widely recognized geologic epoch known as the Anthropocene, wherein planetary processes are marked by human activity. This book is not premised upon a debate over *whether* modern humans are at fault for present environmental conditions. At the outset, we acknowledge and assume responsibility for our ecological predicament.

This collection of essays presents diverse responses to the sombre reality of ecological degradation. We pose a question at the outset: *Given* that we are living in the Anthropocene and are witnesses to the attendant suffering of all forms of life, what changes do we need to make to our lives in order to minimize suffering and maximize well-being for all? Put differently, what does living well, and suffering well, look like in the Anthropocene? These are not simple and easily identifiable changes. The reformation must address our values, structures of meaning, emotional bearings, perceptive matrices, and idiosyncratic tastes, all of which are lodged deep in our unconscious identities, hidden from the light of examination. More complicated still—*knowing* the right thing does not carry easily into *living* the right way. For example, one might register the odious impact of mass consumerism, but the rehabilitation of one's consumer impulses is another matter.

This project of transforming the self, and by extension the culture and community, is well explored in many philosophical and spiritual traditions of antiquity. As autopoietic creatures, we fashion ourselves to adapt to changing inner and outer environments. Granted, not all activities turn out to be ultimately adaptive; what helps us survive in one moment may prove deleterious over a longer course. However, the recognition of our misguided behaviour can compel us to change. We believe that this point has been collectively reached and that we are compelled to make changes to who we are in order to live and suffer well under the increasingly harrowing constraints of the Anthropocene. But what does this project of self-transformation have to do with virtues? And what kind of virtues would qualify as ecological? In an effort to address these questions, this volume presents scholars and practitioners from diverse academic backgrounds who share insights gleaned from vigorous action and thoughtful integrity.

The editors invited authors from a broad spectrum of traditions and disciplines of theory and practice—this in accordance with the first-order ecological principle of *diversity*. These authors represent diversity of academic disciplines, epistemic and methodological orientations, philosophical affiliations, as well as aesthetic visions and writing styles. Each chapter presents creative, thoughtful, and empathic responses. No one discipline, tradition, or orientation has privilege over another. This textual garden facilitates a cross-pollination of thoughts and practices, values and sentiments. From pollen to seed, poetry commingles with philosophy, narrative with argumentation, theory with practice, the political with the personal. And in playing host to multiple genres, we offer an ecological ground for eco-virtues.

Each author reflects on and speaks from lives well examined and well lived, from the spaces of their own research, and from webs of relationality, both personal and professional. These are not disembodied "ivory tower" academicians, for there is no virtue in abstraction. Rather, virtue is the child of *praxis* and *phronesis*, arising from a personal and intimate space that can only be deemed the conscience, or the *soul*. Many readers will read and feel the *soulful-ness* of the text in each unfolding chapter. At the same time, each contribution is academic, marked by intellectual rigour forged at the convergence of incisive scholarship and experiences of lived struggle.

With diversity in mind, we admit that this book is unable to encompass the full range of traditions and disciplines that furnish the annals of human wisdom, nor have we given adequate attention to a range of ecological problems that comprise the present crisis. No single volume can reasonably do so. Missing representations do not indicate oversight on the part of the editors; rather, we were unable to gather such authors under the circumstances, and within the constraints of time. It is our sincere hope that this volume is only the beginning of a larger endeavour, that it may spur others to produce more work of this kind with a greater scope of diversity.

Now, some words about the organization of this volume and about each essay included. Part 1: The Call From and To Earth sets the stage for eco-virtue discourse and invites the reader to consider what it means to practise eco-virtue. Nancy Turner and Darcy Mathews introduce the concepts and practices of "ecosystem services" in their essay "Serving Nature: Completing the Ecosystem Services Circle." Turner and Mathews note that "We are all part of one creation; within it we all have gifts to offer and needs to meet. Our sacredness as earth's humans, animals, plants, and physical features is honoured when both conditions are enacted, requiring our dynamic participation in creation. As humans, as with every order of creation, reciprocal to the gifts from our non-human relations is our responsibility to provide for their needs. It is our role to 'give back' to our non-human relations." They outline human services to nature in the context of Indigenous Peoples' laws and world views with ideas derived from discussions with Indigenous Knowledge Holders, who show how human communities fit into the reciprocal circle of giving and receiving and of virtues reflected in these examples.

In his essay "Ecological Presence as a Virtue," Peter Kahn takes us into the realm of presence, showing that interaction with nature helps people physically and psychologically by providing a portal to a transcendent part of human existence. Kahn argues that we can slow down to create space by being fully present with the more-than-human other. "We would do well to rewild the world and rewild ourselves." Kahn concludes: "Ecological Presence becomes a way of living—a way of being in the world."

Part 2: Morality and Mortality reveals the heart of ecological virtue by plumbing the mystery of life and death. "What could constitute virtue in the circumstances in which we find ourselves?" This is the question Jan Zwicky asks us to consider in her essay "A Ship from Delos." Her answer: "We should approach the coming cataclysm as we ought to have approached life"—with courage, self-control, justice, contemplative practice, and compassion. Our task, in the face of both life and death, Zwicky tells us, is to strengthen our moral fibre.

In the next essay, "Thanatopsis: Death Literacy for the Living," David Greenwood and Margaret McKee ask the reader how considerations of death can be so absent from our ecological attention. They consider the turn toward death a necessary ecological virtue, suggesting that "Talking about death is an act of courage, and listening an act of love: it is saying, 'We are in this together. You are not alone.'" They turn to Whitman's *Leaves of Grass* and offer meditations on his poetry for those "bold enough to open toward the realty of death as a generative force rather than shrink from it in fear of the unknown." They outline a literacy of death to remember its centrality to the ecological dynamics that flow throughout our planet.

Part 3: Insights from the Contemplative Wisdom Traditions takes the reader to the eco-virtues discourse by way of the guidance provided by some of the world's wisdom traditions. Paul Crowe begins his essay on Daoist perspectives on ecological virtues—"What are 'Daoist' Virtues? Seeking an Ethical Perspective on Human Conduct and Ecology"—by pointing out that anyone advocating for change in our relationship to nature faces two daunting challenges: advocating for a fundamental reorientation of our personal ethics, and shifting away from the central concern of ethics within a European and consumerist culture. Crowe is hopeful that, just as human beings have changed their myths and narratives in the past, they can do so again.

In "Never Weary of Gazing: Contemplative Practice and the Cultivation of Ecological Virtue," Douglas Christie draws from the contemplative practices that are part of his Catholic tradition. He refers to Pope Francis's 2014 encyclical *Laudato Si'* in offering a vision of the world as sacramental and as a manifestation of an "integral ecology." He points to a contemplative way of seeing as the pathway to a way of being in and with the world. He writes of such seeing: "[the] sense of being immersed for a time in a world whole unto itself—whole and vibrant and complex and beautiful—and of losing yourself in that world, arrives as a precious gift."

In "The Ecological Virtues of Buddhism," David Loy, using the philosophical foundations of Buddhist teachings, points to the notion of the *bodhisattva*, "a spiritual archetype that offers a new vision of human possibility," as an iteration of virtue. Loy endorses the call for an "eco-sattva" path that "combines personal contemplative practice with an activism that... seeks ways to promote the social and economic changes that are necessary for a more sustainable world."

In Part 4: Philosophies of Virtue Ethics, we present philosophic examinations of virtue in ecology. In "The Ethics of Sustainable Well-Being and Well-Becoming: A Systems Approach to Virtue Ethics," Thomas Falkenberg offers us a derived "ethic of sustainable well-being" in which virtues play an essential role. In the process, he explicates what it means to be alive as a human, develops the concept of *individual* well-being, and connects the latter to socio-ecological systems. In "Why Virtue Is Good for You: The Politics of Ecological Eudaimonism," Mike Hannis argues that concern for our own flourishing as human beings is sufficient to ground a fully formed and workable conception of ecological virtue. Hannis concludes that environmental policies themselves can and should include the sincere promotion and facilitation of ecological virtues.

Part 5: Embodied Creature Connections to Others and Place takes the reader into the realms of the embodied and embedded practices of virtue. In his chapter, "'Owning up to Being an Animal': On the Ecological Virtues of Composure," David Jardine argues that we need to "own up" to being animal, to all the imperfections of our thoughts and deeds, and to "the impermanence and perennial, repeated suffering of the world, human and non-human." To do so, we must compose ourselves, and focus "always and only over the particularities of our lives that we live."

In "Worthy of This Mountain: Living a Life of Friction against the Machine," David Chang recounts his awe-inspiring encounter with Mount Fairweather on the coast of British Columbia. Humbled, he asks himself: "Who am I to behold this majesty? Why should I be here to experience this numinous sight?... How will I live a life that is worthy of this mountain?" By way of a response, he recommends a "life-long engagement in struggle... [an] unremitting vigilance against the infiltration of deleterious power, [and the] injustices that subvert agency and ethical integrity."

In "Stories of Love and Loss: Recommitting to Each Other and the Land," Tommy Akulukjuk, Nigora Erkaeva, Derek Rasmussen, and Rebecca A. Martusewicz each tell a story of love and loss set in very different cultural contexts, wherein they each experienced contradictory "educations"—one formal, based on job or university preparation, the other informal within families, but based on important ecological values. They point to "local virtues" of responsibility, leadership, membership, forbearance, humility, and fidelity as the intangible values offered in families that are needed for just and sustainable communities.

The consummate finale of this collection goes to Carl Leggo and Margaret McKeon's offering, "Evoking Ethos: A Poetic Love Note to Place." The authors encourage a "careful and artful attending to our particular places" as an ecologically virtuous practice and the creation of those places "through our being and doing and in our words, our languages, our stories." They maintain that, "to imagine a changed reality of loving, responsible being, we must first speak our world, in all its magical and terrifying complexity, as if it is a place of love." In the process of writing poetry and poetic creation, the authors argue, "we slow down and linger," adding that we need to hear our own and others' voices "singing out with hopeful courage and conscientious commitment."

We hope that our readers find these essays stirring and moving, conducive to dreaming, envisioning, and activating a forgotten potential. May they inspire our readers to live the ideals of these timeless ecological virtues.

*Part I*

# THE CALL FROM
# *and* TO EARTH

# ONE

# Serving Nature: Completing the Ecosystem Services Circle

NANCY J. TURNER & DARCY MATHEWS

*My grandmother, I watched her when we were little… gathering
medicine or some kind of edible…. [She] didn't go tramping in
the woods and just "chop, chop" or "dig, dig." I could hear my
little grandmother…. She would be chanting a tune as we're
going along…. And when she went up to a tree or a shrub…
she acknowledged [it]…. Like it was a human being…*
—Dr. Mary Thomas, Secwepemc Elder, 2001

## Introduction

A PRIMARY TENANT OF WESTERN SOCIETY IS THAT HUMANS
represent the pinnacle of life, and that through this special status we
can rightfully use any and all of earth's resources for our own ends. This
attitude determines, largely, our behaviours and actions. But this also leads to
certain negative consequences: unprecedented biodiversity loss, deteriorating

water quality, and the ongoing change in the global climate. This statement from a hundred years ago exemplifies this predominant utilitarian perspective:

> All the efforts of the Dominion must be devoted to production and economy. The vast resources of Canada…must be turned to some useful purpose. *Untilled fields, buried minerals or standing forests are of no value except for the wealth which, through industry, can be produced therefrom.*[1]

Not surprisingly, by the turn of twenty-first century, 30 to 50 percent of the planet's surface had been sequestered by humans.[2] This proportion is expanding yearly, as recognized by the Anthropocene age designation. Recognizing the rapid, extensive ecosystem changes caused by humans and the resulting biodiversity losses, the Millennium Ecosystem Assessment (2005) highlighted the "ecosystem services" concept in an attempt to account for some other, indirect value that the earth's ecosystems hold for humans.[3] To this end, four categories of ecosystem services were defined: supporting (nutrient cycling); provisioning (e.g., food, water, genetic, and biochemical resources, medicines, timber, fuel); regulating (of climate, disease, waste treatment, natural hazards, etc.); and cultural (nature's aesthetic, recreational, heritage, spiritual, inspirational, and cultural identity value).

Nevertheless, with this approach nature is still assumed to be the service provider, with us humans as the beneficiaries. Even cultural services represent a one-way flow. Ecosystem services, although an important reminder of our dependence on nature, also fit with dominant society's commodification and monetization of biophysical resources. Many conservationists, environmental philosophers, and ecologists also appreciate nature's intrinsic value.[4] Nevertheless, for many Indigenous Peoples, a major element is still absent in the ecosystem services model.

This missing component is underlain by relationship and reciprocity, embodying an understanding that we humans are an integral part of nature and natural systems, that animals, plants, and even mountains and rivers—considered non-living within Western taxonomies—are our kin, "our relations." We are all part of one creation; within it we all have gifts to offer and needs to meet. Our sacredness as earth's humans, animals, plants, and physical features is honoured when both conditions are enacted, requiring our dynamic participation in creation. As humans, as with every order of creation, reciprocal to the gifts from our non-human relations is our responsibility to provide for their needs. It is our role to "give back" to our non-human relations.

Kwakwaka'wakw Elder Dr. Daisy Sewid-Smith explains:

> We completely depended on nature. The garments that we had, the houses that sheltered us, the foods we ate, the medicines we had—nature supplied it. And that is the reason why we respected nature as we did....Mother Nature does not want you to take from her and not put anything back.... Nature will not survive if we just keep taking![5]

If we honestly see a tree as a relative, a sentient being, we will treat it differently than if we simply think of it instrumentally as "fibre"—an economic object. If we understand that the salmon that provide our food are also our relations with their own needs, their own living spaces, their own families, we will see them and treat them with greater care and appreciation. These recognized responsibilities feature again and again in the teachings, stories, arts, languages, and activities of many the Indigenous Peoples of North America and beyond.[6]

Here, we discuss human services to nature in the context of Indigenous Peoples' laws and world views. We draw on literature reviews and from discussions and interviews conducted over many years with Indigenous Knowledge Holders.

## Giving Back: Human Services to Ecosystems

Indigenous Peoples' systems of traditional ecological knowledge and wisdom reflect a fundamental interconnectedness, both physical and spiritual, between humans and other entities: plants, animals, fish, birds, mountains, water, minerals. Nuu-chah-nulth scholar Umeek (Richard Atleo) explains that this is because all of these are part of creation: "All things are related and interconnected. All things are sacred."[7] Tsilhqot'in witness Gilbert Solomon describes a similar perspective: "When we have to honour all the spirits, we acknowledge the spirits of water, all the elements, the fire, light, the earth, plants, animals. We all believe that they all have spirits, the same spirit that we have, all humans. Spirits are no different."[8] In the Nlaka'pamux creation narratives, "Old-One" instructed the people he had made how to live, to hunt and fish, make baskets, dig roots, and so forth, and also the proper protocols for harvesting plants like saskatoon berry for making a fish spear or yew wood for a bow. Before taking anything from a plant, he taught them, one should "pray or talk nicely to it."[9] Thus, right at humanity's inception, the people learned to treat plants and other living beings with deep respect and to develop relationships with them. Vickers provides a Ts'msyen perspective:

> All—humans, plants, the land, supernatural beings, are alive and intimately related to each other—impacting each other. The continued use of carved and painted crest designs in Northwest Coast art in the form of totem poles, chief screens, ceremonial pieces such as masks and button robes that give account of supernatural encounters or an individual's origins, practical implements such as soapberry spoons, ladles, feast bowls, vests, capes, jackets and shawls remind us of the continued connection between humans, the land and spirit.[10]

For the Rarámuri of northern Mexico, everything that breathes or respires has a soul: plants, animals, humans, stones, the land—all share access to and are moulded by the same air, the same sky, the same water.[11] This relationship is termed "kincentricity," reflecting an understanding that humans and nature are part of an extended ecological family sharing common origins, and acknowledging that a healthy environment is achievable only when humans regard all earth's beings as relatives.[12]

Reconciling and mediating the "kin" versus "resource" roles of non-human life is a key consideration, entailing deep reciprocal relationships and responsibilities to other life forms. Viewing them as sentient beings with intelligence and agency, equally worthy as humans, requires different and more deferential approaches. These beings require our respect, acknowledgement, appreciation, and care, as for our own grandparents or parents. Not only are they our relatives, they are our original teachers—our oldest supporters, provisioners, and helpers.[13]

## The Virtues of Human Services to Nature

This relational attitude toward nature immediately evokes such concepts as responsibility, respect, caring, gratitude, love, and generosity to all other species, just as we would extend these values to our human kin, past, present, and future. Indeed, these are all virtues embedded in diverse Indigenous Peoples' knowledge systems, and in their world views, arts, narratives, ceremonies, and day-to-day activities, as described below.[14]

## Responsibility

Responsibility is about acknowledging relationships. For many Indigenous Peoples it is the responsibility of humans to acknowledge and to look after *all* our relatives, to consider their well-being as inextricably bound to our own.[15]

Kwakwaka'wakw Clan Chief Kwaxsistalla Adam Dick shared the Kwak'wala word *q'waq'wala7owkw*—which means the responsibility of "keeping it living"[16] and reflects accountability in plant cultivation (sowing, tilling, transplanting, weeding, selective harvesting, controlled burning), as well as for animal populations. Other Indigenous Peoples hold similar concepts in ecosystem use and management practices, with humans accountable to the plants and animals they harvest. Responsibility also means dependability, taking these protocols seriously and creating productive gardens or fishing grounds to provide for one's kin, and following the appropriate codes to ensure the continued productivity of habitats and life forms.

## Respect

"Respect is the very core of our traditions, culture and existence."[17] For many Indigenous Peoples, the animal or plant chooses to make itself available to humans, and the success or failure of fishers, hunters, or harvesters depends upon respectful practices.[18] Humans, plants, animals, and spirits are an entangled and intractable whole. As such, reciprocal practices and expressions exist broadly in many Indigenous Peoples' environmental relationships and resource use.[19] For example, the sustainable SX̱OLE reef-net fishery of the W̱SÁNEĆ Coast Salish included a First Salmon ceremony for the first salmon captured.[20] This ritual "underlies the reverent regard for salmon which is one of the principal food animals."[21] Addressing the annual conference of the Association of Professional Biology in 2014, Adam Olsen, a member of the Tsartlip Nation and of the Legislative Assembly of British Columbia, describes the respectful relationship each fisherman must have with salmon this way:

The wealth of a Straits Salish reef netter was not counted by the number of sockeye he could catch, preserve, and trade alone. … [It] was also calculated by the long-term quality and abundance of the fishing grounds he owned, cared for and passed down to his descendants. …This relationship was central to the way of life of a reef netter. Rather than viewing the sockeye as a resource to be exploited in full, reef netters believe the fish, like every other living being, was once a human … [and that] all living things were an integral part of the circle of life. The sockeye that passed by our nets were honoured as sacred lineages, no different than the lineage of reef netters … both were part of the same cycle.[22]

Conservation and sustainability are implicit in this reciprocal practise of gratitude. That ethos also encompasses a strict rejection of disrespectful acts—overhunting, overharvesting, or the unwarranted harming of animals—and extends to respect for that which had already been harvested through injunctions against such things as playing with one's food or with parts of a dead animal. Transgressions are understood to precipitate bad luck or bad weather events.[23]

## Gratitude

Indigenous Peoples worldwide, despite their differences, are rooted in cultures of gratitude arising from the tenet of respect.[24] Showing appreciation and respect to everything obtained or used was expressed in the quote with which this chapter was introduced. Mary Thomas's mother, Christine Allen, whenever receiving even a small gift of berries or other food, would hold them up and say to the Creator, "Kwukstsámcw, Kwukstámcw, Kwukstsámcw!" (Thank you, thank you, thank you!).[25] For many, this expression of gratitude is threaded throughout daily life, in modest but significant rituals underscoring people's relational connections with other species.

These everyday acts complement more formal communal rituals, such as highly prescribed First Foods ceremonies honouring the seasonal arrival of important foods. These were consistent in form, with acknowledgement of the food species as "Friend, Supernatural One," and thanking it for providing itself to people. Often, it is asked to protect against illness or bad fortune. Harvesting and consuming key plants and animals were ritually enacted, acknowledging and celebrating their value to those entrusted as their stewards, and ensuring that each key food species would return in subsequent years.[26]

## Caring

Upholding the well-being of the world and its human, plant, animal, and spirit inhabitants is exemplified by the prohibitions, threaded through various Indigenous world views, against greed, waste, and selfishness in resource use

Devil's-Club Man. *Gitkinjuaas (formerly Giitsxaa, Ron Wilson), Haida artist (Turner 2004)*

and management.[27] Medicinal plants have special harvest protocols, such as showing gratitude to the plants and ensuring that medicines are never wasted. Since medicinal plants are strongly spiritual, all prepared medicines have to be used up, or bad luck will ensue.[28]

*Love*

Environmental health and sustainability—cornerstones of human health and well-being—require individuals and communities to care about their environment, to love their home places, to feel belonging with the people, plants, and animals who are their family and community.[29] Mary Thomas, mentioned previously, outlined the concept of place as a locale of love and gratitude— for family, community, and nature—exemplified by a long-ago family trip to Mount Revelstoke, now a Canadian national park:

> I can remember as a little girl running, hopping, skipping, jumping through all these beautiful flowers—that's one of the happy memories that I have.... The children were taught to respect Mother Nature and to appreciate it, and when you breathe in this cool air and you can imagine yourself sleeping out here in open air—we just had a little lean-to, and you're breathing in this beautiful mountain air.... Even now you can smell that *melanllp* [subalpine fir], from the beautiful boughs.... And every time you smell that beautiful smell of Mother Nature's creation, you appreciate it, you love it...you become a part of it.[30]

Personal connections to and relationships with nature motivate people to care about the future, and not just their own future but others' as well.[31] Mary Thomas explained, "We were connected to Mother Nature, we were not superior. We are a part of Mother Nature. If we destroy Mother Nature, we are destroying ourselves."[32] Humans have an emotional and ontological need for connection with nature. When children encounter nature, they build the cognitive constructs necessary for emotional and intellectual development and life-long learning.[33] "Nature-deficit disorder" is a label used today to address the physical, intellectual, and emotional cost to children increasingly deprived of direct contact with nature.[34] Time in nature promotes learning about seasonal and biological cycles and ultimately about how to sustain and support communities.[35] For those people with a deep love of place, knowledge and learning are "situated" and contextualized, encompassing aesthetic, ceremonial, economic, personal, familial, historical, political, and practical considerations.[36]

## Generosity

The ethics of generosity and responsibility are widespread throughout Indigenous society. Leaders having proprietorship over specific productive areas such as fishing grounds not only hold rights of access but also the responsibility for the care and maintenance of these places, and for ensuring their ultimate benefit to the entire community.[37] For the Coast Salish, the basis for chiefly prestige—"being of good name"—was not just a matter of birth but also of exemplifying generosity and responsibility.[38] This ethos extended to people of all ages and statuses. Young girls were taught to give away the first basket of berries they picked, and young boys to share their first salmon caught or first deer hunted with their Elders or those in need. Through such learning opportunities, young people cultivated generosity and empathy toward others, creating an endless intergenerational support system fuelled simultaneously by the receiver's appreciation and the giver's pride and satisfaction.

# Indigenous Approaches to Environmental Relationships

The virtues of responsibility, respect, gratitude, caring, love, and generosity are exemplified in three kinds of ecological management practised by coastal BC Indigenous Peoples: clam garden mariculture, western redcedar harvesting, and blue camas cultivation. Each involves management and focused care with reciprocal benefit to ecological health and biodiversity. Colonial misperceptions or misrecognition of Indigenous ecological knowledge have made us largely blind to the kinds and extent of such management practices, and they have also skewed our sense of the numbers of Indigenous Peoples living here before European contact and the catastrophic epidemics following Europeans' arrival in the New World. Prior to contact, however, Indigenous villages—some very large—had occupied most bays, inlets, and other inhabitable parts of the coastline. In many places, clam gardens, western redcedar stands, blue camas prairies, and myriad other managed habitats are ubiquitous and still evident.

## Clam Garden Mariculture

Clams—especially butter clams and cockles—have for millennia provided the coastal peoples of British Columbia with dependable, easily harvested food. A vitamin-rich source of protein, clams could be eaten fresh almost year-round, dried and stored for later use or trade, and served at community events and gatherings. Clam use likely commenced as soon as postglacial sea levels began to stabilize and intertidal shellfish populations developed. Origin stories, rituals, songs, and archaeological evidence underscore the long-term economic and cultural significance of clams, with shell midden sites on the Northwest Coast appearing around 6,400 years ago.[39]

For at least the past millennium, however, Northwest Coast Peoples have not just harvested clams, but have actively and expertly managed clam habitat by clearing large stones from existing beds and building intertidal rock wall terraces.[40] Termed "clam gardens" or "sea gardens," these features are known in Kwak'wala as lúxwxiwey (or "rolled rocks to clear an area"): a term pertaining to their creation and maintenance.[41] Other groups have similar names; the Nuu-chah-nulth term is t'iimiik, or "something being thrown" or "move aside rocks,"[42] and in Tla'amin (Coast Salish), wúxwuthin means "rocks piled on beach" or "like a breakwater."[43]

Clam gardens concentrate shellfish resources in accessible locations. They range from restricted push-outs of cleared stones, to monumental rock walls or terraces stretching hundreds of metres, and they allow for selective harvesting and increase the numbers, quality, and possibly size of the clams.[44]

Furthermore, these rock walls provide habitat for crabs, sea cucumbers, octopuses and other invertebrates, as well as smaller fishes, and anchorage for edible seaweeds.[45] In effect, they were sites of perpetual harvesting for a diversity of resources at favourable tides, but they need to be monitored, tended, and harvested to maintain healthy habitats—a form of reciprocity for the gift of the clams.

## Western Redcedar Harvest Trees

First Peoples' relationship with cedar is so profound that it is hard to envision the culture and history of the Pacific Northwest without western redcedar (*Thuja plicata*), the "tree of life."[46] Integral to the lives of the region's

TOP: Clam gardens on the Central Coast of British Columbia, near present-day Bella Bella. BOTTOM: Rock walls built to create new clam habitat. *Darcy Mathews*

First Peoples for at least five thousand years,[47] the tree's importance is reflected in a proliferation of heavy wood-working tools, waterlogged cedar basketry and other artifacts, and the ubiquity of culturally modified cedars in the archaeological record.[48]

People collected the bark from living trees in spring and summer, either in long, tapering strips or in rectan-gular sheets that were used for the roofing and siding of houses. Harvesters addressed

Three-thousand-year-old wet-site-preserved burden basket from the Esquimalt Lagoon, southern Vancouver Island.[49] Made from western redcedar wood slats and split withes and manufactured by a wrap-around plaiting method (visible in white square). *Darcy Mathews*

the tree's spirit with respect, explaining why the bark was needed, and thanking the tree for its provisions.[50] Tall, straight trees with few branches were best to harvest bark from. Bark of yellow cedar—or *Xanthocyparis nootkatensis*—was harvested and used similarly. Cedar wood was, and is, used for building con-struction, canoes, totem and monumental poles, bentwood storage, and cooking boxes, and is integral to food storage and economic and social exchange.[51]

Cedar withes, strong and flexible, were split and used for heavy-duty rope and lashings and openwork burden baskets. The roots were also harvested and used to make beautiful yet utilitarian baskets of a variety of styles and sizes, with distinctive overlaid patterns from bitter cherry bark and grass stalks.[52] People recognized that humans and trees are linked in a dynamic reciprocity of giv-ing, taking, and transformation. A standing tree from which house planks were removed is called *keto'q* ("begged from") in Kwak'wala, and before obtaining boards in this way, one should look upward to the tree and say, "We have come to beg a piece of you today. Please! We hope you will let us have a piece of you."[53]

Gratitude for cedar is exemplified in this Kwakwaka'wakw prayer, delivered by a woman harvester of cedar bark:

> Look at me, friend! I come to ask for your dress, for you have come to take
> pity on us; for there is nothing for which you cannot be used...for you are

Examples of two western redcedar harvest trees on the Central Coast of British Columbia, including rectangular bark sheets (above, with ecologist Kira M. Hoffman) and house planks (following page) removed from living, standing trees. *Darcy Mathews*

really willing to give us your dress. I come to beg you for this, Long-life Maker, for I am going to make a basket for lily roots out of you. I pray, friend, not to feel angry with me on account of what I am going to do to you and I beg you, friend, to tell our friends about what I ask of you. Take care, friend! Keep sickness away from me, so that I may not be killed by sickness or in war, O friend![54]

Given its multiple values, slow growth rate, and intensive use, there is always a risk of over-exploiting cedar. Harvesting has the potential to kill or seriously damage a tree. To overcome this risk, ecological ethics—namely, responsibility and respect—are implicit in First Peoples' management practices relating to cedar use and to forests in general.[55] There are, in fact, indications that bark harvesting "primed" cedar trees for future bark harvesting. The lobes that develop when tree bark or planks are harvested compartmentalize and heal over, and bark from them was often subject to later bark harvesting, perhaps because thinner-walled cambial cells may have had comparable qualities to the bark of younger trees.[56] A tree with multiple bark harvest scars likely produced several times more bark over its lifetime than if all the bark had been removed at one time.[57] Thus, keeping the tree alive through small-scale, frequent harvesting actually allowed a greater yield, at the same time fulfilling ethical and spiritual responsibilities to the tree and its surrounding ecosystem.

Objects made with western redcedar bark, withes, and wood are "pieces of places,"[58] materials collected in one place, brought to another, and transformed into capes, canoes, and memorial poles. In this process, each item is thus connected to the larger landscape. Furthermore, the living, modified trees from which these materials were harvested—evidencing long tapering or rectangular bark strip scars or plank removal scars—are "figures in the landscape"[59] that serve as mnemonic and visual cues to a deep history of inhabitation and the management of forest resources, linking localized village life with the broader and entangled landscapes of fjords, islands, and mountainous slopes.

### Blue Camas Cultivation

For centuries, the Coast Salish Peoples of coastal southwestern British Columbia engaged in the long-term, sustainable production and management of blue camas (*Camassia leichtlinii* and *C. quamash*), called ḴȽO,EL (qʷłaʔl) in Straits Salish (see top figure on page 19). ḰȽO,EL's edible bulbs were the "number one vegetable," the most nutritionally and economically important root food of the Coast Salish.[60] A "cultural keystone species,"[61] camas has played a defining role in the economies, social hierarchies, cultural identities, and land use practices of the Lekwungen, W̱SÁNEĆ (Saanich), Hul'qumi'num, and other Coast Salish Peoples. Often referred to as the "Salmon People," the Vancouver Island Salish could just as accurately be called the "Camas People."[62] A major source of carbohydrates, camas was associated with complex management activities and systems of production, preservation, exchange, and redistribution.[63] The long-term pit-cooking process (depicted in bottom figure on page 19) allowed the chemical breakdown of inulin, its largely indigestible major carbohydrate, into easily digested, sweet-tasting fructose and fructans.

Families owned camas-digging areas and tended these intergenerationally through clearing, weeding, and controlled burning.[64] Intensified but selective harvesting and processing of camas may relate to population pressure and increased sedentism in the centuries prior to European contact.[65] In any case, maintaining camas prairies through fire and clearing, ownership of patches by individuals who oversaw tending and harvesting of camas, selection of bulbs by size and depth, their use as a valued trade item, and stories and ceremonies relating to camas all signify a careful, respectful, appreciative relationship with camas and its locales, extending well back in time. Significantly, camas prairies and their maintenance seem to be inextricably related to burial cairns on southern Vancouver Island, indicating a close relationship between camas production and human ancestors.[66]

## Making a Difference for the Future

The ecological virtues of responsibility, respect, gratitude, care, love, and generosity contributed to the sustainability and resilience of plants, animals, and places through time. These virtues are reflected in multiple, complex ways, rendering the presence of people and their tending and harvesting practices as beneficial to forest ecology and tree health in ways not widely recognized. For example, the large quantities of clams and other shellfish harvested by Northwest Coast First Peoples (from clam gardens, at least in the last thousand years) accumulating in midden sites have served as calcium reservoirs enhancing forest

TOP: Dense remnants of cultivated blue camas persist today. *Darcy Mathews*
BOTTOM: Large archaeological earth oven (about 4.5 metres in diameter) that was likely used to cook camas bulbs, associated with an almost three-thousand-year-old Coast Salish inhabitation near the Esquimalt Lagoon. *Kristi Bowie*

growth and have also created flat, well-drained house platforms for people. Furthermore, the calcium-rich anthropogenic soils at these sites have directly benefitted cedars, as higher wood calcium produced greater radial growth and decreased top die-back.[67] As such, the clustering of bark- and wood-harvesting cedar trees around village sites may not just be a matter of accessibility, but also a prime example of reciprocity among humans, clams, and cedar.

These three examples of traditional management systems serve as lessons for all of us as we struggle to balance our need for resources and sustenance with our harmful impacts on other species and on the earth's life-support systems. Regarding our relationships as *reciprocal*[68]—not just assuming "ecosystem services for humans" but considering "human services and responsibilities to ecosystems and to other species"—is one of these lessons. Maintaining an attitude of respect and appreciation goes hand in hand with engendering a sense of responsibility and caretaking. As Gitxsan Hereditary Chief Delgamuuxw declared in his address to the BC Supreme Court in 1987: "The land, the plants, the animals and the people all have spirit—they all must be shown respect. That is the basis of our law."[69] Kwakwaka'wakw cultural specialist Kim Recalma-Clutesi expressed similar convictions: "We believe that all life, whether it be animal, plant or marine is sacred and is as important as human life. We are but one small part of the big picture and our food gathering practices and ceremonies remind us of that."[70] The results of taking responsibility and caring for other species and their well-being are generally positive in terms of productivity and quality of the resources they provide for us.[71]

It is essential for us humans to recognize our complete dependence on the natural world—to understand that our economy is grounded in the environment and cannot take precedence over it or even be given equal consideration.[72] How can we convey this truth effectively to those who would ignore it? Transformational education is required.[73] Conveying these virtues in terms of our relationships to other species and the environment, through teachings, stories, participatory learning, and living examples—common means of imparting knowledge in Indigenous societies—would help mainstream society to come to an understanding and shift its attitudes and actions accordingly.[74] Just like friendships and relationships that flourish and deepen the more people invest in them, and in turn enrich the lives of those who give, our relationships with nature enrich us in multiple and diverse ways as we make the effort to develop them.

In the spirit of the recommendations recently put forward by the Truth and Reconciliation Commission of Canada in its Calls to Action,[75] Canadians are asked to adopt and implement the UN Declaration on the Rights of Indigenous Peoples, to which we are signatories. The declaration stresses, in article 25.1, the rights of Indigenous Peoples everywhere "to maintain and strengthen their distinctive spiritual relationship with their traditionally owned or otherwise occupied and used lands, territories, waters and coastal seas and other resources and to uphold their responsibilities to future generations in this regard." This includes recognizing Indigenous Peoples' ties to their lands and their relationships with other species.

The concepts of spirituality, kincentricity, and the interconnectedness of all things must have ancient origins, extending back in some form to the "beginning of time" since they are firmly embedded in the origin stories of many Indigenous families and clans, as well as in ceremonies and other traditions. Despite many decades of cultural repression, Indigenous Peoples have still retained strong and enduring cultures, including their deep sense of responsibility to their lands and their non-human relatives.[76] In some cases these bonds have been strengthened in recent years as cultural renewal and revitalization eclipse decades of cultural suppression. Language revitalization is part of this cultural strengthening, so that words reflecting relationality and respect are again heard in ceremonial contexts. For example, the ceremonial names that the SENĆOŦEN (Straits Salish) give to the different kinds of salmon, expressed during the First Salmon ceremony and other ceremonies, are still used by some: the sockeye is addressed as "Honoured One"; the coho as "Parent of Your Daughter- or Son-in-Law"; the humpback, or pink salmon, as "Adopted Niece or Nephew"; and the spring, or chinook, as "Sister- or Brother-in-Law."[77] These names themselves imply people's sense of appreciation and responsibility to the salmon. Indeed, as Earl Claxton Sr. and John Elliott Sr. have written, "All these ceremonies were performed in respect to the salmon, our relative, who sacrifices his life so that we may continue to live."[78] Similarly, the Kwak'wala words spoken during the Sacred Cedar Ceremony of the Kwakwaka'wakw carry forward the deep responsibility and gratitude people hold for this tree and all it represents.[79]

Often the very people who hold and value these "kin" relationships and reciprocity world views are not consulted when planning, decision-making, and policy formulation around use of nature are undertaken. Although many non-Indigenous people also recognize our responsibilities to the land and to other species, there is a need for many more in mainstream society—especially planners and decision-makers—to embrace this view. In the long run, a more selfish view, focused on the accelerating accumulations of wealth through increasingly unsustainable exploitation of the earth and its species without considering the needs of all, will make all of us poorer.

The use of ancient stories and ceremonies as conservation tools may well be rejected out of hand by ecologists and land managers, since they do not fit the model for direct interventions implied by some definitions of the term "conservation." Yet they may be far more effective in communicating complex concepts and key principles, in promoting effective learning and understanding, and in motivating people to participate in implementing more sustainable strategies for long-term resource use than approaches based on scientific knowledge alone. Collectively, they gently instil and effectively reinforce the codes of proper behaviour, consideration, and restraint required to live well in a kincentric world.[80]

## Conclusions

The ecological crisis we find ourselves in today, with the ongoing spectre of climate change and the accelerating extinctions of other life forms, puts into perspective the limits of humanness as a privileged category: a result of the intellectual tradition of Western modernity, in which nature only exists as a category in contrast to culture and reason. This world view, alien to many of the Indigenous Peoples of the world, is too simplistic; it cannot sustain us into the future. We need to draw on the depths, experiences, and virtues reflected in *all* the world's societies in order to teach ourselves to limit our negative impacts on our environment, on other species, and on our own lives. In the words of Pueblo educator Gregory Cajete, "Western society must once again become nature-centered, if it is to make the kind of life-serving, ecologically sustainable transformations required in the next decades."[81] Making an effort to understand and embrace different perspectives, including those of Indigenous Peoples, to look beyond the assumption of human supremacy, and to realize that we have a need to care for nature and other species, can make all the difference.

## Acknowledgements

We are grateful to Gitkinjuaas (formerly Giitsxaa) for permission to use the "Devil's-Club Man" image and to Kristi Bowie for permission to use the "earth oven" image. We are grateful as well to the many wise, kind, and generous Indigenous Knowledge Holders who have explained their cultures, languages, and perspectives to us and others over many years. These include, among so many others: Dr. Richard Atleo (Umeek), Dr. Luschiim Arvid Charlie, Dr. Gregory Cajete, Elsie Claxton, Dr. Nicholas XEMŦOLTW Claxton, Helen Clifton, Clan Chief Adam Dick (Kwaxsistalla), Roy Haiyupis, Chief Ernie Hill, Dr. Robin Kimmerer, Dennis Martinez, Dr. Aaron Mills, Joan Morris (Sellemah), Dr. Melissa Nelson, Adam Olsen, Kim Recalma-Clutesi (Oqwilowgwa), Dr. Daisy Sewid-Smith (Mayanilth), Linda Smith, and Dr. Mary Thomas. We would also like to thank Drs. Heesoon Bai, David Chang, and Charles Scott for inviting us to contribute to this volume and for the work they are doing to make the world a better place. Our thanks, as well, to copyeditor Ryan Perks and the rest of the publishing team at University of Regina Press.

## Notes

1   H.N. Whitford and R.D. Craig, *Forests of British Columbia* (Ottawa: Commission of Conservation, 1918), 1; emphasis added.

2   See P.M. Vitouset, H.A. Mooney, J. Lubchenco, and J.M. Melillo, "Human Domination of Earth's Ecosystems," *Science* 277 (1997): 494–9.

3   See A.D. Guerry et al., "Natural Capital and Ecosystems Services Informing Decisions: From Promise to Practice," *Proceedings of the National Academy of Sciences* 112, no. 24 (2015): 7348–55; C.M. Raymond et al., "Ecosystem Services and Beyond: Using Multiple Metaphors to Understand Human–Environment Relationships," *BioScience* 63 (2013): 536–46.

4   B.S. Reyers et al., "Finding Common Ground for Biodiversity and Ecosystem Services," *BioScience* 62 (2012): 503–7; A. Drengson and Y. Inoue, *The Deep Ecology Movement: An Introductory Anthology* (Berkeley: North Atlantic Books, 1995); Aldo Leopold, *A Sand County Almanac, with Essays on Conservation from Round River* (1949; New York: Ballantine, 1990).

5   D. Sewid-Smith, address delivered at Helping the Land Heal conference, University of Victoria, November 1998.

6  F. Brown and K. Brown, with B. Wilson, P. Waterfall, and G. Cranmer Webster, *Staying the Course, Staying Alive: Coastal First Nations Fundamental Truths* (Victoria, BC: Biodiversity BC, 2009); M. Siisip Geniusz, *Plants Have So Much to Give Us, All We Have to Do Is Ask: Anishinaabe Botanical Teachings* (Minneapolis: University of Minnesota Press, 2015); R. Kimmerer, *Braiding Sweetgrass, Indigenous Wisdom: Scientific Knowledge and the Teachings of Plants* (Minneapolis: Milkweed Editions, 2013); N.J. Turner, *The Earth's Blanket: Traditional Teachings for Sustainable Living* (Vancouver: Douglas & McIntyre; Seattle: University of Washington Press, 2005); N.J. Turner, *Ancient Pathways, Ancestral Knowledge: Ethnobotany and Ecological Wisdom of Indigenous Peoples of Northwestern North America*, vols. 1 and 2 (Montreal: McGill–Queen's University Press, 2014).

7  Clayoquot Scientific Panel, *First Nations' Perspectives on Forest Practices in Clayoquot Sound*, Report 3 (Victoria, BC: Cortex Consulting and Government of BC, 1995), 15. See also E.R. Atleo (Chief Umeek), *Principles of Tsawalk: An Indigenous Approach to Global Crisis* (Vancouver: UBC Press, 2011), and Atleo, *Tsawalk: A Nuu-chah-nulth Worldview* (Vancouver: UBC Press, 2004).

8  Supreme Court of British Columbia, "*Tsilhqot'in Nation v. British Columbia,* 2007 BCSC 1700: Reasons for Judgement," p. 132, https://www.bccourts.ca/Jdb-txt/sc/07/17/2007BCSC1700.pdf.

9  J.A. Teit, *Mythology of the Thompson Indians*, vol. 7, part 2: *The Jesup North Pacific Expedition*, in F. Boas, ed., *Memoirs of the American Museum of Natural History* (New York: G.E. Stechert, 1912).

10  P.J. Vickers, "Ayaawx (Ts'msyen Ancestral Law): The Power of Transformation" (PhD diss., University of Victoria, 2008), 67.

11  E. Salmón, "Kincentric Ecology: Indigenous Perceptions of the Human-Nature Relationship," *Ecological Applications* 10, no. 5 (2000): 1327–32.

12  R., Senos, F. Lake, N. Turner, and D. Martinez, "Traditional Ecological Knowledge and Restoration Practice in the Pacific Northwest," in *Encyclopedia for Restoration of Pacific Northwest Ecosystems*, ed. Dean Apostol, 393–426 (Washington, DC: Island Press, 2006). See also Turner, *The Earth's Blanket*.

13  Kimmerer, *Braiding Sweetgrass*; Turner, *Ancient Pathways, Ancestral Knowledge*, vols. 1 and 2.

14  Kimmerer, *Braiding Sweetgrass*; Turner, *Ancient Pathways, Ancestral Knowledge*, vols. 1 and 2.

15  Turner, *Ancient Pathways, Ancestral Knowledge*, vol. 1: 310. See also Turner, *The Earth's Blanket*.

16  D.E. Deur and N.J. Turner, "Introduction: Reconstructing Indigenous Resource Management, Reconstructing the History of An Idea," *Keeping It Living: Traditions of Plant Use and Cultivation on the Northwest Coast of North America* (Seattle: University of Washington Press; Vancouver: UBC Press, 2005).

17 Haiyupis in Clayoquot Scientific Panel, *First Nations' Perspectives on Forest Practices in Clayoquot Sound*, 6–7.

18 F. Berkes, *Sacred Ecology: Traditional Ecological Knowledge and Resource Management*, 3rd ed. (Philadelphia: Taylor and Francis, 2012), 107.

19 N.J. Turner and F. Berkes, "Developing Resource Management and Conservation," *Human Ecology* 34 (2006): 475–8.

20 N.X. Claxton, "To Fish as Formerly: A Resurgent Journey back to the Saanich Reef Net Fishery" (PhD diss., University of Victoria, 2015).

21 E. Gunther, "An Analysis of the First Salmon Ceremony," *American Anthropologist* 28 (1926): 617.

22 A. Olsen, Speech to the Association of Professional Biology, Victoria, BC, April 26, 2014.

23 For example, G.T. Emmons, *The Tlingit Indians* (Seattle: University of Washington Press, 1991); N.J. Turner, M.B. Ignace, and R. Ignace, "Traditional Ecological Knowledge and Wisdom of Aboriginal Peoples in British Columbia," *Ecological Applications* 10 (2000): 1275–87.

24 Kimmerer, *Braiding Sweetgrass*, 106.

25 Turner, *The Earth's Blanket*.

26 F. Boas, *Ethnology of the Kwakiutl, 35th Annual Report, Parts 1 and 2* (Washington, DC: Smithsonian Institution, Bureau of American Ethnology, 1921); Turner, *Ancient Pathways, Ancestral Knowledge*, vol. 2: 329.

27 Turner and Berkes, "Developing Resource Management and Conservation"; Turner, Ignace, and Ignace, "Traditional Ecological Knowledge and Wisdom of Aboriginal Peoples in British Columbia."

28 Turner, *Ancient Pathways, Ancestral Knowledge*, vol. 1: 327, 463.

29 E.N. Anderson, *Caring for Place: Ecology, Ideology, and Emotion in Traditional Landscape Management* (Walnut Creek, CA: Left Coast Press, 2014), 18. See also Turner, *Ancient Pathways, Ancestral Knowledge*, vol. 1: 403.

30 M. Thomas, N.J. Turner, and A. Garibaldi, " 'Everything Is Deteriorating': Environmental and Cultural Loss in Secwepemc Territory," in *Secwepemc People and Plants: Research Papers in Shuswap Ethnobotany*, ed. M.B. Ignace, N.J. Turner, and S.L. Peacock, 366–401 (Tacoma, WA: Society of Ethnobiology, 2016), 385.

31 Turner, *Ancient Pathways, Ancestral Knowledge*, vol. 1: 403.

32 Thomas, Turner, and Garibaldi, " 'Everything Is Deteriorating,'" 381.

33 M. Driessnack, "Children and Nature-Deficit Disorder," *Journal for Specialists in Pediatric Nursing* 14 (2009): 73–5.

34 R. Louv, *Last Child in the Woods: Saving Our Children from Nature-Deficit Disorder* (Chapel Hill, NC: Algonquin Books, 2008).

35 Turner, *Ancient Pathways, Ancestral Knowledge*, vol. 1: 403.

36  S. Feld and K. Basso, eds., *Senses of Place* (Sante Fe, NM: School of American Research Press, 1996); V.D. Nazarea, "Lenses and Latitudes in Landscapes and Lifescapes," in *Ethnoecology: Situated Knowledge/Located Lives*, ed. V.D. Nazarea, 91–105 (Tucson: University of Arizona Press, 1999).

37  N.J. Turner, R.Y. Smith, and J.T. Jones, "A Fine Line between Two Nations: Ownership Patterns for Plant Resources among Northwest Coast Indigenous Peoples—Implications for Plant Conservation and Management," in Deur and Turner, *Keeping It Living*, 151–80.

38  W. Suttles, *Economic Life of the Coast Salish of Haro and Rosario Straits* (New York: Garland, 1974), 495.

39  K.M. Ames and H.D.G. Maschner, *Peoples of the Northwest Coast: Their Archaeology and Prehistory* (London: Thames and Hudson, 1999), 88–9.

40  M.E. Caldwell et al., "A Bird's Eye View of Northern Coast Salish Intertidal Resource Management Features, Southern British Columbia, Canada," *Journal of Island and Coastal Archaeology* 7, no. 2 (2012): 219–33; D. Deur, A. Dick, K. Recalma-Clutesi, and N.J. Turner, "Kwakwaka'wakw 'Clam Gardens,' " *Human Ecology* 43 (2015): 201–12; A.S. Groesbeck et al., "Ancient Clam Gardens Increased Shell Fish Production: Adaptive Strategies from the Past Can Inform Food Security Today." *PLoS One* 9 (2014): 1–13; D. Lepofsky and M. Caldwell, "Indigenous Marine Resource Management on the Northwest Coast of North America," *Ecological Processes* 2, no. 1 (2013): 12; D. Lepofsky et al., "Ancient Shellfish Mariculture on the Northwest Coast of North America," *American Antiquity* 80 (2015): 236–59; J. Williams, *Clam Gardens: Aboriginal Mariculture on Canada's West Coast* (Vancouver, BC: New Star Books, 2006).

41  D. Deur et al., "Kwakwaka'wakw 'Clam Gardens' "; D.L. Mathews and N.J. Turner, "Ocean Cultures: Northwest Coast Ecosystems and Indigenous Management Systems," in *Conservation for the Anthropocene Ocean: Interdisciplinary Science in Support of Nature and People*, ed. P.S. Levin and M.R. Poe, 169–99 (San Diego: Academic Press, 2017).

42  R. Bouchard and D. Kennedy, *Clayoquot Sound Indian Land Use* (Report prepared for MacMillan Bloedel Limited, Fletcher Challenge Canada, and British Columbia Ministry of Forests, 1990), 386.

43  D.I.D. Kennedy and R. Bouchard, *Sliammon Life, Sliammon Lands* (Vancouver, BC: Talonbooks, 1983), 147–8.

44  D. Deur et al., "Kwakwaka'wakw 'Clam Gardens' "; Groesbeck et al., "Ancient Clam Gardens Increased Shell Fish Production."

45  Caldwell et al., "A Bird's Eye View of Northern Coast Salish Intertidal Resource Management Features."

46  H. Stewart, *Cedar: Tree of Life to the Northwest Coast Indians* (Vancouver: Douglas & McIntyre, 1984).

47 R.J. Hebda and R.W. Mathewes, "Holocene History of Cedar and Native Indian Cultures of the North American Pacific Coast," *Science* 225 (1984): 711–13.

48 Ames and Maschner, *Peoples of the Northwest Coast*; D. Lepofsky and N. Lyons, "The Secret Past Life of Plants: Paleoethnobotany in British Columbia," *BC Studies* 179 (2013): 39–84; A.H. Stryd and M. Eldridge, "CMT Archaeology in British Columbia: The Meares Island Studies," *BC Studies* 99 (1993): 184–234.

49 D. Mathews and P. Dady, *Data Recovery and Monitoring at DcRu-74: Colwood Trunk Sewer Line Installation, Permit 2003-402* (Report on file at the British Columbia Archaeology Branch, 2003).

50 Stewart, *Cedar: Tree of Life*, 113.

51 A. Cannon and D.Y. Yang, "Early Storage and Sedentism on the Pacific Northwest Coast: Ancient DNA Analysis of Salmon Remains from Namu, British Columbia," *American Antiquity* 71 (2006): 123–40.

52 D.R. Croes, "Northwest Coast Wet-Site Artifacts: A Key to Understanding," in *Emerging from the Mist: Studies in Northwest Coast Culture History*, ed. R.G. Matson, G. Coupland, and Q. Mackie, 51–75 (Vancouver: UBC Press, 2003), 73–4.

53 E.S. Curtis, *The North American Indian*, vol. 10: *The Kwakiutl* (Evanston, IL: Northwestern University Library, 1915), available at http://curtis.library. northwestern.edu/curtis/toc.cgi.11. See also Boas, *Ethnology of the Kwakiutl*, 616–17.

54 Boas, *Ethnology of the Kwakiutl*, 619.

55 N.J. Turner, D. Deur, and D. Lepofsky, "Plant Management Systems of British Columbia's First Peoples," *BC Studies* 179 (2013): 107–33.

56 M. Eldridge and A. Eldridge, *The Newcastle Block of Culturally Modified Trees* (Report on file at the BC Archaeology Branch, 1988), 44, cited in J. Earnshaw, "Cultural Forests of the Southern Nuu-chah-nulth: Historical Ecology and Salvage Archaeology on Vancouver Island's West Coast" (MA thesis, University of Victoria, 2016).

57 Earnshaw, "Cultural Forests of the Southern Nuu-chah-nulth," 168.

58 Sensu R. Bradley, *An Archaeology of Natural Places* (London: Routledge, 2000).

59 Sensu C. Bierwert, *Brushed by Cedar, Living by the River: Coast Salish Figures of Power* (Tucson: University of Arizona Press, 1999).

60 B.R. Beckwith, " 'The Queen Root of This Clime': Ethnoecological Investigations of Blue Camas (*Camassia leichtlinii* (Baker) Wats., *C. quamash* (Pursh) Greene; Liliaceae) and its Landscapes on Southern Vancouver Island, British Columbia" (Victoria, BC: Department of Biology, University of Victoria, 2004); Suttles, *Economic Life of the Coast Salish of Haro and Rosario Straits*; N.J. Turner and H.V. Kuhnlein, "Camas (*Camassia* spp.) and Riceroot (*Fritillaria* spp.): Two Liliaceous 'Root' Foods of the Northwest Coast Indians," *Ecology of Food and Nutrition* 13 (1983): 199–219; N.J. Turner and R.J. Hebda, *Saanich Ethnobotany: Culturally Important Plants of the WSÁNEC People* (Victoria: Royal BC Museum, 2012).

61  A. Garibaldi, and N.J. Turner, "Cultural Keystone Species: Implications for Ecological Conservation and Restoration," *Ecology and Society* 9, no. 3 (2004), http://www.ecologyandsociety.org/vol9/iss3/art1.

62  J. Lutz, *Makuk: A New History of Aboriginal-White Relations* (Vancouver: UBC Press, 2008), 66.

63  Beckwith, "'The Queen Root of This Clime,'" 85.

64  N.J. Turner, " 'Time to Burn': Traditional Use of Fire to Enhance Resource Production by Aboriginal Peoples in British Columbia," In *Indians, Fire and the Land in the Pacific Northwest*, ed. R. Boyd, 185–218 (Corvallis: Oregon State University Press, 1999).

65  A.V. Thoms, "The Northern Roots of Hunter-Gatherer Intensification: Camas and the Pacific Northwest" (PhD diss., Washington State University, 1989).

66  D.L. Mathews, "Funerary Ritual, Ancestral Presence, and the Rocky Point Ways of Death" (PhD diss., University of Victoria, 2014), http://hdl.handle.net/1828/5637.

67  A.J. Trant et al., "Intertidal Resource Use over Millennia Enhances Forest Productivity," *Nature Communications* 7, 12491 (2016), doi:10.1038/ncomms12491.

68  R. Trosper, *Resilience, Reciprocity and Ecological Economics: Northwest Coast Sustainability* (London: Routledge, 2009).

69  G. Wa and D. Uukw, *The Spirit in the Land: The Opening Statement of the Gitksan and Wet'suwet'en Hereditary Chiefs in the Supreme Court of British Columbia, May 11, 1987* (Gabriola Island, BC: Reflections, 1989), 7.

70  Kim Recalma-Clutesi, personal communication with Nancy J. Turner, 1992.

71  Brown and Brown, *Staying the Course, Staying Alive*.

72  K.M.A. Chan et al., "Opinion. Why Protect Nature? Rethinking Values and the Environment," *Proceedings of the National Academy of Sciences of the United States of America (PNAS)* 113, no. 6 (2016): 1462–5, doi:10.1073/pnas.1525002113; T. Satterfield, R. Gregory, S. Klain, M. Roberts, and K.M. Chan, "Culture, Intangibles and Metrics in Environmental Management," *Journal of Environmental Management* 117 (2013): 103–14.

73  B.R. Beckwith, N.J. Turner, and T. Halber Suarez, " 'You have to do it': Creating Agency for Environmental Sustainability through Experiential Education," in *Routledge Handbook of Environmental Anthropology*, ed. H. Kopnina and E. Shoreman-Ouimet, 412–27 (Oxford: Routledge, 2017); G. Raygorodetsky, *The Archipelago of Hope: Wisdom and Resilience from the Edge of Climate Change* (New York: Pegasus Books, 2017); G. Snively and Wanosts'a7 L. Williams, eds., *"Knowing Home": Braiding Indigenous Science with Western Science* (Victoria: University of Victoria, 2016), https://pressbooks.bccampus.ca/knowinghome/.

74  Turner and Berkes, "Developing Resource Management and Conservation."

75 See especially Call to Action 43, available at "Truth and Reconciliation Commission of Canada: Calls to Action," http://trc.ca/assets/pdf/Calls_to_Action_English2.pdf.

76 Turner, *The Earth's Blanket*.

77 E. Claxton Sr. (YELḰÁTȻE) and J. Elliott Sr. (STOLȻEȽ), *Reef Net Technology of the Saltwater People* (Brentwood Bay, BC: Saanich Indian School Board, 1994).

78 Claxton and Elliott, *Reef Net Technology of the Saltwater People*, 27–35.

79 D. Sewid-Smith (Mayanilth) and Clan Chief A. Dick (Kwaxsistalla), interviewed by N.J. Turner in "The Sacred Cedar Tree of the Kwakwaka'wakw People," in *Stars Above, Earth Below: Native Americans and Nature*, ed. M. Bol, 189–209 (Pittsburgh: Carnegie Museum of Natural History, 1998).

80 Atleo, *Principles of Tsawalk*.

81 G. Cajete, *Native Science: Natural Laws and Interdependence* (Santa Fe, NM: Clear Light, 1999), 266.

TWO

# Ecological Presence
# as a Virtue

PETER H. KAHN, JR.

## Preamble

I WAS FIFTEEN YEARS OLD AT THE TIME, ON A MONTH-LONG MOUN-
taineering course, west of the Grand Tetons, on a high mountain plateau.
Our instructor pulled us together as a group. He said, "Here's the plan.
You'll be bivouacking tonight. You can use your clothing. But you can't use
your sleeping bag or tent. The point of this is to give you a sense of what a real
bivouac in the mountains would be like. Have a good night. We'll see you in
the morning."

I retrieved all of my clothing from my backpack. I had wool trousers, a wool
sweater, a wool jacket, a rain jacket, a hat, socks, and gloves. I found a spot on
some rocks overlooking a canyon, with the Tetons further to the east. There
were about twelve of us students. We were all looking for and finding nice spots.
The night came on. It got dark. I sat there. It got colder. I sat there. Mountain
air. Cold stars above. I tried lying down, but it was not so comfortable and I

grew colder from the rocks. I pulled my clothing tighter around me. Actually, I was having a very good time. I thought, "I'm really doing it! I'm bivouacking!" After a while it got colder again, and there wasn't anything more I could do. I just sat there and went a little numb. Time slowed down. I couldn't hurry it; it just *was*. For some periods I must have dozed. Later into the night, a crescent moon rose in the east and provided some gentle light to the canyon below and to the snow-capped mountains. My mind slowed down, but there were times when it was very alert. Stillness. It went on. Then, finally, from that spot I began to notice ever so slowly the sky awakening, and I watched the sun rise as if it were the first morning in Eden. In hindsight, all these years later, I marvel at my youth. To have experienced such purity and innocence, and to be aware of it at the time and yet simultaneously not aware, at least not in this sense that all of us have of looking back on our youth and setting it in the relief of a lifetime—it is like young love. It is a benediction on life.

In this chapter I would like to put forward a new term: Ecological Presence. It will be a little odd to use words to talk about Ecological Presence because at its core it is not about words or ideas, theories or evidence. Ecological Presence is an experience of perceptions that can emerge through interaction with nature, wherein those perceptions can then settle into a mind's awareness without conscious mind activity. Sometimes nature can lead you pretty directly into Ecological Presence. You have likely experienced it in your life —maybe often. I had a glimmer of it during my bivouac.

## The Ontology and Epistemology of Presence

As humans, we are degrading, polluting, and destroying nature at an astonishing rate, distancing ourselves from what remains. In response, lines of scientific research have emerged, and at times coalesced, showing that people need to interact with nature to do well, both physically and psychologically.[1] The research literature shows, for example, that interaction with nature can reduce stress, depression, aggression, crime, and ADHD symptoms; it can improve immune function, eyesight, and mental health, and increase people's social connectedness.[2] There is also research in positive psychology on how interaction with nature contributes to the pleasant life, good life, and meaningful life.[3]

These are important lines of research because most of us live with a Western, modernist world view that privileges scientific knowledge, and if science says that human-nature interaction is important, then that provides leverage for many agendas that seek to conserve and enhance nature and people's interactions with it. But at their core, these approaches not only treat nature as yet another resource to be consumed for human benefit but also assume that the empirical world is the only domain that human consciousness can apprehend, and is thus worthy of systematic inquiry. Other world views provide a wider perspective on what we in Western thought traditions call "ontology" and "epistemology."

Ontology refers to the nature of being, existence, and reality, and of categories therein. Ontological questions include, for example, some of the following: What exists? What is real? How can we describe what is real? Are there basic categories of being, such as "alive" and "not alive," and of time, number, and space? What does it mean for something to have existence? In turn, epistemology takes any truth claim, especially ontological truth claims, and asks, How do we know? The modern scientific response to such questions is to assert (or seek to establish) that what exists, ontologically speaking, is objectively "out there" in the physical universe, and, epistemologically speaking, is established through hypothesis-based studies, with measurable outcomes that can be replicated by independent people.

There are, however, other responses. One of the basic ontological moves of such responses is to assert (or seek to establish) that there are dimensions of reality that exist beyond the physical world. Some of these responses draw on thousands of years of Asian or Indian metaphysics. For example, about twenty-five hundred years ago the Chinese philosopher Lao Tzu wrote in the *Tao Te Ching* about the *Tao*, that nameless quality that exists without physical existence (form) and yet underlies and makes form possible:

> Tao is empty
> > yet it fills every vessel with endless supply
> Tao is hidden
> > yet it shines in every corner of the universe.[4]

How can something be empty and yet full? Hidden yet emergent? By analogy, if you are in a room right now, look around and see what is there. As I look around my study, I see a desk, bookcases, books, a mirror, a light fixture, and several photographs. But what is missing in such a description is the empty space that makes it all possible. If there were no empty space in my study, could anything in this study exist at all? Using a similar analogy, Lao Tzu writes:

> Clay is molded to form a cup
>   yet only the space within
>   allows the cup to hold water.[5]

Thus, something is ontologically constituted by what it is—its form—and what it is not—its empty space. That is the *Tao*.

We could also use the word "Being" in place of *Tao*. For something to *be*, it has to have both form and no form. A human Being can also be understood in these terms. As a Being, we are ontologically constituted by these two dimensions. The dimension of form includes not only our physical body but also all of our thoughts and emotions. The dimension of no-form is that of emptiness. Eckert Tolle calls it Presence.[6] In Buddhism, it is called, Nothingness (Śūnyatā), or the Great Void.

A term such as "Nothingness" is hard to grasp. To grasp it is to lose it because it is not a thing: its ontological properties make it resistant to concepts and language. This idea frames the very opening to the *Tao Te Ching*:

> The Tao that can be spoken of is not the Tao itself.[7]

Of course, Lao Tzu then proceeds to speak of the *Tao* through all of his verses! He obviously believes there is something of value that can be said about it. At its core, his basic message is: do not confuse the words about the *Tao* with the *Tao* itself.

Words, in this sense, can be thought of as a signpost in the woods. Imagine hiking for a few days toward a destination, call it Big River. You find yourself disoriented at some point. You then come upon a signpost that says, "Big River, 3 miles," with an arrow pointing to the left. The signpost is helping you get oriented, but it is not to be confused with the entity it is pointing toward. If you were thirsty at the signpost, it would not provide you with the water that lies at Big River. Likewise, we can talk about Nothingness, Space, and Being. But all the words of the world are not the thing. No form of the world is the thing because Being is no-thing.

There is a story of a Zen master, Gutei, who was often asked questions about Zen.[8] You can imagine some of these were short and sweet, such as, "Master, what is Zen?" That is an ontological question. Some questions likely combined ontology and epistemology, such as, "Master, how do we know you know what Zen is?" Whatever the question, Gutei was known to always answer it the same way: by simply raising his finger. That answer is not so satisfying to the human mind.

This story then goes on to say that a boy, a young student of Gutei's, began to imitate the master's answer. Whenever anyone asked the boy what Zen was, the boy raised his finger. When Gutei heard about this, he grabbed hold of the boy and cut off his finger. The boy cried out in horror and began to run away. At that moment, Gutei called to him to stop. The boy stopped, turned around, and looked at Gutei. Gutei raised his finger. In that instant the boy was enlightened. There: is that more satisfying to the human mind?

The story is both entertaining and hard to decipher, as most Zen stories are. One of the commentaries on it emphasizes that neither Gutei's nor the boy's enlightenment had anything to do with the finger. If one believed they did, one was hopelessly deluded; perhaps as deluded as one who tried to drink water from a signpost that reads, "Big River."

That said, Nothingness, Emptiness or Formlessness, Space, Being, Śūnyatā, Zen, or the *Tao* that "is empty yet it fills every vessel with endless supply"— whatever the name—can emerge in human consciousness. But we know it is not by virtue of raising a finger or cutting off a finger. How, then?

Another Zen story offers a suggestion. Student goes to Zen master and says that he will be leaving the monastery but would greatly appreciate if the master could write down detailed instructions for Zen, so that he has them while he is gone. The student gives the master a piece of paper and a brush. The master writes, and then hands the paper back to the student. The student is expecting some profound writing but all he reads is the word "Attention." The student says, "Surely, master, there is much more to Zen, can't you please write a little more?" The master takes back the paper and brush, writes some more, and then hands it back to the student. The student then reads a second word: "Attention." The student is feeling a little frustrated, believing that surely the master is holding back vital knowledge. He presses one more time, and one more time the student reads the word "Attention." All told: "Attention. Attention. Attention."

Being has the potential to become known when ordinary compulsive and conditioned thinking stops and the mind is highly attentive. The emergent awareness is not what happens when we go below thought, as when we sleep or "zone out" watching television or playing video games. It is not the going below consciousness that happens when we have a couple of drinks, and then perhaps a third or a fourth. For in those states, one is aware of mostly nothing, but not Nothing.

With these ideas as the intellectual grounding, I can now begin to situate what I mean by Presence. Presence is that awareness of Nothing, Space, and Being that emerges in human consciousness when compulsive conditioned thinking stops and the mind is attentive.

## Is Presence Mumbo-Jumbo?

If you have not experienced something like Presence in your life, then much of this essay so far may well seem nonsensical. That is fine. You would have some good company. Freud, for example, had trouble fathoming the experience that a friend of his had expressed as an "oceanic feeling." Freud had sent this friend his book *The Future of an Illusion*, wherein he (Freud) argues that religions espouse false dogmas that people accept because of infantile wishes. Freud then reports that his friend

> answered that he entirely agreed with my judgment upon religion, but that he was sorry I had not properly appreciated the true source of religious sentiments. This, he says, consists in a peculiar feeling, which he himself is never without, which he finds confirmed by many others, and which he may suppose is present in millions of people. It is a feeling which he would like to call a sensation of "eternity," a feeling as of something limitless, unbounded—as it were, "oceanic." This feeling, he adds, is a purely subjective fact, not an article of faith; it brings with it no assurance of personal immortality, but it is the source of the religious energy which is seized upon by the various Churches and religious systems, directed by them into particular channels, and doubtless also exhausted by them.[9]

Freud goes on to say: "The views expressed by the friend whom I so much honour…caused me no small difficulty. I cannot discover this 'oceanic' feeling in myself."[10] And there is the rub. If you have not experienced the oceanic feeling—a similar term for Presence, Being, or the *Tao*—how can you believe it as anything but dogma or illusion, or both?

There is a cabin in the mountains of Northern California that I built. I spend my summers there, writing and living close with the land. The cabin is off the grid, with a few solar panels for electricity. There is an outhouse with a view of big mountains. There is a hot shower under an enormous oak tree. During the heat of the summer, wasps fly through the open back door into the cabin and then try to leave through the plate glass windows in the front. They cannot get out that way. They spend all day flying around that glass window trying to get out. They die trying. At the day's end, I sweep up dozens of dead wasps. I have tried to help them by "explaining" to them that there is an open window next to the plate glass window, which I specifically keep open for them, and if they flew just five feet over they would exit the cabin and be well and live free. But their brains are not constituted to understand human language, or even rudimentary human truths that perceive how physical objects spatially exist in relation to other objects.

Like a wasp, we are also a product of biological evolution, with capabilities and limitations. It seems to me patently evident that the human brain is constituted to understand some but not all of what exists in the empirical world. As a case in point, consider that traditional Polynesian sailors once navigated thousands of miles of open ocean by the stars, sky, and the currents and swells of the ocean, guided by observation, interpretation, and intuition.[11] We modern people are not able to do so. This difference seems to indicate that humans are far more capable of understanding the world than we give our brains credit for.

According to Einstein's relativity theory, if we travelled out into space at just about the speed of light for a year, then turned around and came home at the same speed, we would be two years older while everyone else would be about fourteen thousand years older. Time is not what it seems. Indeed, as early as 1908, Hermann Minkowski said that time is connected to space. Einstein described it as a space-time continuum that becomes distorted by objects. Or try to imagine another astronomical idea: that our universe came into being 14 billion years ago through the Big Bang, which originated from a single entity smaller than an atom—that was actually about "a million billion billion times smaller than a single atom."[12] Can you imagine anything that small? Can you imagine our universe that small? To pose a spatial ontological question: What was on the "other side" or "outside" of that infinitesimally small particle? Or an

ontological time question: What existed before the Big Bang? Some astrophysicists posit a "two-headed time" theory whereby time travelled backward to the
Big Bang and is now travelling forward in time. Other astrophysicists posit that
currently multiple universes exist simultaneously.

Using language established earlier, I would say that the ideas and theories
that come from astrophysics are remote signposts to a reality beyond what we
think of as the "real" world. The questions we ask—such as, "What existed
before the Big Bang?"—make sense to us because of the way our brains are
constituted through human evolution on this planet. But surely these questions cannot be answered, at least not in the way our minds ask them. It is a
little like a wasp asking, "Can I get through the plate glass window at its top or
at its bottom?" The answer is neither.

I would suggest that it is entirely rational that a rationalist would agree that
we cannot rationally understand everything, or perhaps even much—not in a
deep sense—about the universe. But, if you are the skeptical rationalist, then
I would like to invite you to keep an open mind, and not immediately seek
a reductionistic account of Presence. Such a reductionism occurs when it is
proposed, for example, that the experience of Presence is an illusion based on
wishes from childhood (as Freud did), simply an odd by-product of human
evolution, or an epiphenomenon of human mental processing.

## Ecological Presence

In the wilderness area some twenty miles away from my cabin, one chooses
summer routes carefully with attention to water. When I was there last August,
I had hiked up over the summit of a mountain and dropped down onto the
other side to where a small spring shows on a map. But when I got there, the
spring was just damp mud. It took me a few hours, until nightfall, to reach
another spot I knew of with a trickle coming out of the earth. Later on that trip,
I left a river valley and set out cross-country up a desolate ridge that some years
earlier had been burnt to a crisp in big fires. Since then this area had regrown
with endless tall thickets of thorn bushes. As I was hiking upward, I sought
with each step to find both good footing and a way through the thorns, as much
as that was possible. I could smell the burned, fallen trees, and often had to
climb over them. My body was smudged with charcoal. It was hot. I knew, as I

sought to cross from one mountain drainage to another, I needed to be pretty good with my route-finding because I only had two litres of water for the full day. As I was climbing, I sensed a slight hint of green up a distant slope. I set course for that. When I got there I saw that the green vegetation was some ten feet tall. I dropped my pack, poked around the vegetation, and then pushed my body into it. And there I found a little spring of water. It was roughly a hand deep and a hand wide, coming out of that harsh ridge. I drank my fill. It was an oasis. It was a spot of time that stretched forever, even as I was dripping sweat and swatting bugs.

Interaction with nature has a way of helping us to be attentive, to stop the endless mental chatter, and thus affords us the opportunity to experience Presence in nature, with nature, and through nature: that is another way to characterize Ecological Presence.

Three ancestral ways of interacting with nature tend to engender Ecological Presence. In our evolutionary history, these forms were often seamlessly connected. For illustrative purposes, consider a method of hunting used by the Ju/wasi bushman of the Kalahari Desert in the early 1950s. This method presumably mirrored a way of hunting that existed in our evolutionary history.[13] On a hot day, with temperatures hovering around 120 degrees Fahrenheit, the hunter, with spear in hand, would find and then begin to run toward a bull eland. The bull eland could run fast, and would run some distance away. The hunter would run after him. As the now resting bull eland saw the hunter beginning to get close, he would run away again. The hunter would keep pursuing. The eland would run again. This pattern repeated itself for an hour. Two hours. Four hours. It could continue for five hours or more. At some point, the bull eland would overheat and collapse from exhaustion. The hunter would then be face to face with an exhausted but still dangerous wild animal.

Up to this point in the hunt, the hunter's emotions are pretty even-keeled, with a uniform expenditure of energy. There was alertness while tracking the bull eland. The hunter stayed aware of his surroundings; aware of potential animals that could be harmful to him; aware of his own strengths; aware of the weather; and aware of his location. It was a calm but very focused awareness that lasted for hours and emerged through the steady stream of his physical engagement with his natural landscape.

But at the moment of his direct encounter with the collapsed bull eland, the hunter's physically engaged calmness changes dramatically. Highly alert, the hunter assesses the eland's condition. Is it possible that the eland could quickly rise and charge him? His senses seek and provide relevant information. His mind synthesizes it, mostly intuitively. There is no formal calculus that says "the sands are two inches deep at this spot, which slows my normal striking stride by 43 percent; bull eland has been running 5.63 hours, which in this heat translates to his being at 6 percent efficiency; temperature of 118 degrees Fahrenheit has led to my active pulse rate of 90 beats per minute, which means that…" No, that is not how it works. Rather, through Ecological Presence the understanding of these facts emerge. The hunter raises his spear. He balances it in his hand perfectly. He chooses when, or it could be said he lets the choice choose him, because there is now no difference between these two statements, and he takes three long, quick steps toward the bull eland, just as close as presence tells him to, and immediately thrusts his spear deep into the animal's midsection. As it hits its mark, he just as quickly scampers some additional feet away because a wounded animal is unpredictable and most dangerous. Adrenaline courses through the hunter's body. The bull eland dies right there in the desert sand.

It is a peak experience. Primal aggression. The hunter has killed big life so that he and his people can live. His genetic programming makes him optimally geared—physically, intellectually, and emotionally—to engage in killing like this. The hunter is in the moment, highly conscious: ecologically present. Killing requires his full attention. For this form of Presence, he does not need a Zen monk scribbling on a piece of paper the words "Attention. Attention. Attention." He has learned this lesson through nature. Attend or be unsuccessful in the hunt. Attend or be killed by the animal you are hunting. Attend or die.

But now here is the thing with the hunter's heightened experience of the kill: He does not keep enacting it. He has killed the eland. There is nothing more to kill. What happens next? Well, the hunter disembowels the animal, and quarters it in ways that allow him to carry it. That takes time. And then there is the long, slow walk home, to his camp, to his people, with the immense weight on his back. The adrenaline that had surged in his body during the kill is now gently absorbed. This could be said to be part of nature's natural system of stress reduction, which becomes even more restorative once the hunter arrives at camp. There, he is greeted by community and family, and he feels safety. Dusk settles. Fires are lit. The meat begins to roast over the coals. Night comes on. There is meat for everyone in his small tribe. The hunter rests around the

fire ring, eating, talking, filling his belly, replenishing with liquids, sharing stories of the day, of the hunt. He lies back and looks up at the night sky: a million
stars overhead. Sleep comes easy.

According to the biophilia hypothesis,[14] we have adapted and evolved from
our ancestral days, and still have affordances and constraints, dispositions,
loves, and fears that are the product of our ancestral selves. From this perspective, it is possible to see expressions of the hunter's experience of the kill in
some modern-day activities. For one thing, over 16 million people go hunting
each year in the United States. Most of these people do not require the meat
from the animal to sustain their lives, and yet they go anyway, which indicates
a legacy of this desire to hunt and to experience the kill. Less directly, could it
be said that we see vestiges of the ancestral hunt when people today engage in
dangerous, adrenaline-producing physical activity in nature, such as climbing
thousand-foot vertical walls in Yosemite, parachuting out of airplanes, or kay-
aking raging rivers? Perhaps vestiges can been seen in competitive sports, like
football: a receiver catches a pass and a linebacker, running full speed, times his
hit perfectly, smashing the receiver to the ground; it is almost like he is using
his body as a spear. Millions of people watch such sports and live this experience vicariously. Indeed, ancestral parts of ourselves, when distanced too
far from healthy expression in nature, may find expression in increasingly distorted and sometimes addictive pursuits. There are people, for example, who
spend large chunks of a day engaged in violent video games, killing in a digital
world. It could be said that this part of our ancestral brain allows people to get
addicted to stimulant drugs, such as amphetamines and cocaine, wanting more
and more of the "action," not being able to say no.

Constraints and affordances of our genetic code give new meaning to
Shakespeare's dictum, "What's past is prologue."

I have conveyed three ancestral means for how human-nature interaction
can foster Ecological Presence. One is through even-keeled physical exertion
in a wild landscape. A second is by dealing with an immediate challenge and/or
dangerous natural situation. A third is through gentle restorative immersion in
nature. These ancestral means are diminished today because of our destruction
of nature and our control over it. However, in principle they are still available to
our modern minds, certainly in partial forms: by means, for example, of hiking
through old-growth forest, birding in urban wetlands, swimming in big ocean
waves, encountering a coyote, or lying in sunshine on a beach. Such forms of

human-nature interaction offer a portal into certain forms of Being. They can open up some Space in human consciousness.

Yet these ancestral means are not the end of what human consciousness is capable of ascertaining in terms of Ecological Presence. They are our beginnings. By means of an analogy, consider a radio telescope—a dish that collects radio waves from distant galaxies and converts the waves into electrical signals, which astronomers then analyze. The first one was developed by an engineer, Karl Jansky, with Bell Telephone Laboratories in 1932; was able to receive shortwave radio signals at a frequency of 20.5 megahertz. From this radio telescope, Jansky identified radio signals from three sources: nearby thunderstorms, distant thunderstorms, and what he eventually deduced was the Milky Way.[15] Since then, these radio telescopes have continued to increase in size and sophistication. The largest one today (completed in 2016) is 500 metres in diameter with an area the size of thirty football fields, built into a hillside in China. They need to be pretty big because radio waves are extremely weak. "A cell phone signal is a billion times more powerful than the cosmic waves our telescopes detect."[16]

Modern mind emerged from Paleolithic mind. Where, when, and why is another story, and one not easily told.[17] We may justifiably speculate that our minds today, in comparison to Paleolithic minds, have developed to be more creative, experimental, inquisitive, driven toward novelty and innovation, and to have greater flexibility and more degrees of freedom. These qualities may provide our modern minds with the potential to access Presence more directly and deeply, as modern radio telescopes provide access to radio signals that originate from farther and farther reaches of the universe.

That said, our modern minds face certain limitations in accessing Ecological Presence. For one thing, in comparison to Paleolithic interactions with nature, we have nowhere near the depth of relation with nature and wild nature that helps engender Ecological Presence. For another, our modern minds are often engaged in an endless churning of thought after thought after thought, as if thinking can save us from our thoughts—if we just think harder or more, which our minds cannot.

I remember one mountaineering trip when I attempted a solo climb on the West Buttress of Denali. It is a popular route, but you can still die on it. On the lower part of the mountain, I pulled 60 pounds of gear in a sled up the glacier, while carrying another 60 pounds on my back. I felt like a pack mule on skis. While trudging upward, my head would be pointed mostly down. But whenever I rested and looked around, I would begin to see—if the weather was clear—the Alaska Range emerge, and then step by step fall beneath me.

At 11,000 feet, with crampons and ice axe on the now steeper terrain, I split the load, and did carries to establish myself at 14,000 feet. The hard work of carrying heavy loads to this point seemed to have had two effects on me. One was that the seemingly endless commotion of my mind eased up over the hours and days. Maybe that is because there was hardly any new social stimuli coming into my mind, and because the challenging environment required my attention in the moment. Another effect, likely because of the first, was that my psyche opened up to the mountain. I felt that I was a small part of it, with it, strong in body and mind and yet but a speck in that vast white landscape, humbled, as one who lies in a remote forest meadow, and, before slumber, looks up at the endless stars overheard, a bounty of awe, galaxies beyond galaxies—who can fathom?

I waited out some storms at 14,000 feet, and then chose my day to head up a 1,000-foot headwall onto the West Buttress, comprising one of the classic lines on the mountain—an ascending ridgeline, sometimes only a few feet wide. It was an unusual day in that I had the route basically to myself. At one point, I had intestinal troubles, but the terrain was too steep for me to do anything, so I continued upward like that, holding, holding, holding, and then finally I found a spot level enough, dove my ice axe into the icy snow, kicked out a little spot for my feet, and unzipped my down pants. After thirty seconds of intestinal expulsion, I saw that I had made a potentially enormous mistake, as my ice axe was no longer tethered to my harness or wrist, but standing free in the icy snow—I had placed it there because I was desperate to get out of my clothes before I defecated in them. The ice axe started to come out and fall the 2,000 feet directly below. But I grabbed at it—I had only one shot for it—and somehow brought it back to me. It was my lifesaver. I cannot believe I made that mistake. I clipped the ice axe back into my harness, and I held it to me.

In other situations—such as in running a marathon—you can cross over the edge of your limit and live to learn from it. For example, if you go out too fast in a marathon, much faster than you had trained, you might cramp up in pain at mile twenty-two and have to drop out of the race. That is not so bad; you will get home that afternoon. But if you are solo on a big mountain and go past your limit, that is bad. There is no backup. You will die up there, or die falling. And the thing with getting close to the edge of one's limit is that there is no answer book that can tell you where exactly that edge is.

I pulled my pants back up. I sipped a little tea that was still warm in my thermos. Then I was surprised to feel that my stomach was all right. I thought, "Okay. Wow, I'm okay to keep heading up." I regrouped.

There are three potential points of contact when climbing like this: two feet with crampons, and an ice axe in hand. For safety, the technique in climbing is always to maintain two points of contact. I reminded myself of that: "Ice axe and crampon in the ice and the other foot steps up. Breathe. Move ice axe up. Step. Breathe. Step. Move ice axe up. Step. Breathe. Step. Move ice axe up." It had started happening earlier that day, but at this point I became aware of it directly: all of my consciousness was focused on the immediate demands of the moment, even as all of my consciousness was becoming aware of another dimension of Presence that lay just behind this physical one. I was living two lives at once. I kept track of time. It mattered. But in another sense time did not move like it usually did. It felt stationary, like a vast ocean, as if my movement created the flow of time itself. At 17,200 feet, at a plateau on the ridge that formed the high camp on this route, I willed my legs past and upward, but they would not respond. I have always believed that one's body can do more than one's mind says it can. But at that point, I knew that I was not going higher.

I rested in the snow. I needed to regroup again. I needed to draw on new strengths to get myself down. I reminded myself that most accidents happen on the descent. Downward I went. I repeated a sort of mantra: "I'm strong. I'm paying attention. I'm strong. I'm paying attention." I would like not to fall to my death. The clarity of Presence now mostly left me, and my mind had partly shut down. I was not balancing well. Hours later I got to the top of the headwall. I dropped off the ridgeline. The first part of the headwall had a fixed line, and I was relieved when I clipped into it. I was more relieved later when I clipped out and awkwardly jumped over a *bergschrund*—a crevasse formed by the glacier separating from the ice on the headwall. I saw my tent in the basin below. If I fell now, I could stop myself easily. I let myself look with satisfaction on the mountains and glaciers cascading below me. By 10:30 p.m. I was only minutes from my tent. A member of another climbing group was outside his. He watched me approach. He asked if I was all right. I said, "Yes!" I was flooded with feelings of safety. He said that I was walking very slowly, weaving from side to side. He said it looked like I was drunk and double-checked that I was okay. I thought I had been walking this last part fast and steady. Perhaps I was a little closer to my edge than I had known.

For an experienced mountaineer, this is not usually a difficult route to summit. But that is a statement that compares one to another. What I find more interesting is how wild nature, at whatever level of physical activity one can bring forward, can bring one into Presence. When I am an old man, frail in body, I would like to be able to open my door and walk the five minutes outward away from the safety of my cabin, across a little stream, through a few feet of forest, to a meadow that looks out onto the mountains of my youth. Maybe I will need a cane. Maybe I will need a walker. Maybe I will look like a drunk weaving from side to side. But through it, I hope to feel the Presence that is at the centre of us all.

## Postamble

The way forward? I would like to suggest that we build on our Paleolithic mind, which is still, in some deep ways, a part of us. For that, we would do well to bring into our lives the three forms of human-nature interaction described above: daily physical exertion in a natural, if not completely wild, landscape; dealing with natural challenges and, occasionally, somewhat threatening natural situations; and daily gentle, restorative immersion in nature. We would do well to rewild the world and rewild ourselves.[18]

With over 7.5 billion people on our planet, and with more than half of the world now living in urban areas, is such rewilding really feasible? In some ways, yes, at least incrementally.[19] Indeed, I have been working in this area for some years, as my colleagues and I seek to develop a design methodology—what I call interaction pattern design—that provides the means to deepening and rewilding, not just nature, but the interactions that people have with nature.[20]

What is an interaction pattern? Think about a meaningful way that you have interacted with nature, and then characterize it in such a way that you could see the same thing happening with different forms of nature. That is an interaction pattern. For example, it is wonderful to walk along the edge of a lake or along a river. The pattern could be named as "walking along the edges of water." Notice that this pattern, once named, can be enacted in countless different environments. Walking along the seashore or walking around a lake in the city. It can be enacted walking alongside a public fountain that is designed to allow for this form of interaction. Each enactment of an interaction pattern is different and can embody attributes that are more urban or more wild. Each pattern, however, shares a common feature—namely that you can easily recognize the pattern whenever it occurs. Knowing this, every form of interaction can be made slightly more wild. By understanding that rewilding helps lead to Being and Presence, interaction pattern design then becomes a practical and visionary way forward.

I remember some years ago attending a talk at my university by a Buddhist monk. He said that our outer environment is not so important; nor does it really matter what happens with our technological systems. What matters, he said, is where our consciousness is. Afterward I said to him that I understood it mattered where our consciousness was, but given that we have physical form, and retain biological affordances and constraints from our evolutionary history, doesn't it merit our sincere efforts to shape our built environment and protect our natural environment to help foster more aware and present consciousness?

It is in this sense that Ecological Presence is a virtue. It is part of a theory not of what is right—as in right and wrong, which guides traditional analytical moral theorizing, often focusing on consequentialist and deontological arguments.[21] Rather, Ecological Presence is aligned more with a theory of the Good, with roots stretching back to Aristotle and the *Nicomachean Ethics*. However, I would say that Ecological Presence is most aligned with Eastern traditions that deepen it. Ecological Presence becomes a way of living—a way of being in the world. Nature helps us to recalibrate our modern minds by minimizing thinking when it is redundant or counterproductive, which it often is. When we live with Space in our consciousness, then when we think, our thoughts are more often original and of a higher calibre. Compassion and love become not efforts but natural expressions. The human brain, through mind, has the capacity to integrate the form and the formless, and to perceive with endless depth into an ontology of Presence, living within it, through it, as an elevation of life itself. We do that, and we will redefine our species.

## Notes

1   T. Hartig et al., "Nature and Health," *Annual Review of Public Health* 35 (2014): 207–28, doi:10.1146/annurev-publhealth-032013-182443; H. Frumkin et al.,

"Nature Contact and Human Health: A Research Agenda," *Environmental Health Perspectives* 125, no. 7 (2017): doi:10.1289/EHP1663; P.H. Kahn, Jr., *Technological Nature: Adaptation and the Future of Human Life* (Cambridge, MA: MIT Press, 2011); E.O. Wilson, *Biophilia* (Cambridge, MA: Harvard University Press, 1984).

2 R. Berto, "The Role of Nature in Coping with Psycho-Physiological Stress: A Literature Review on Restorativeness," *Behavioral Sciences* 4, no. 4 (2014): 394–409; M.S. Taylor et al., "Research Note: Urban Street Tree Density and Antidepressant Prescription Rates—A Cross-Sectional Study in London, UK," *Landscape and Urban Planning* 136 (2015): 174–9; D. Younan et al., "Environmental Determinants of Aggression in Adolescents: Role of Urban Neighborhood Greenspace," *Journal of the American Academy of Child & Adolescent Psychiatry* 55, no. 7 (2016): 591–601; F.E. Kuo and W.C. Sullivan, "Environment and Crime in the Inner City: Does Vegetation Reduce Crime?" *Environment & Behavior* 33, no. 3 (2001): 343–67; F. Kuo and A. Faber Taylor, "A Potential Natural Treatment For Attention-Deficit/Hyperactivity Disorder: Evidence from a National Study," *American Journal of Public Health* 94 (2004): 1580–6; G.A. Rook, "Regulation of the Immune System by Biodiversity from the Natural Environment: An Ecosystem Service Essential to Health," *Proceedings of the National Academy of Sciences* 110, no. 46 (2013): 18360–7; M. He et al., "Effect of Time Spent Outdoors at School on the Development of Myopia among Children in China: A Randomized Clinical Trial," *JAMA* 314, no. 11 (2015): 1142–8; G.N. Bratman, J.P. Hamilton, and G.C. Daily, "The Impacts of Nature Experience on Human Cognitive Function and Mental Health," *Annals of the New York Academy of Sciences* 1249, no. 1 (2012): 118–36; M.T. Holtan, S.L. Dieterlen, and W.C. Sullivan, "Social Life under Cover: Tree Canopy and Social Capital in Baltimore, Maryland," *Environment & Behavior* 47, no. 5 (2014): 502–25.

3 M.E.P. Seligman, *Authentic Happiness: Using the New Positive Psychology to Realize Your Potential for Lasting Fulfillment* (New York: Free Press, 2002).

4 L. Tzu, *Tao Te Ching*, trans. and commentary J. Star (New York: Penguin, 2001), verse 4, p. 17.

5 Tzu, *Tao Te Ching*, verse 11, p. 24.

6 E. Tolle, *The Power of Now* (Novato, CA: New World Library, 2004).

7 L. Tzu, *Tao: A New Way of Thinking*, trans. and commentary C. Chung-yuan (New York: Perennial, 1977), verse 1, p. 1.

8 P. Reps, *Zen Flesh, Zen Bones: A Collection of Zen and Pre-Zen Writings* (1957; repr., North Clarendon, VT: Tuttle Publishing, 1985).

9 S. Freud, *Civilization and Its Discontents*, trans. and ed. James Strachey (New York: W.W. Norton, 1930/1962), 11.

10 Freud, *Civilization and Its Discontents*, 11.

11  W. Davis, *The Wayfinders: Why Ancient Wisdom Matters in the Modern World* (Toronto: House of Anansi, 2009).

12  N. Szokan, "What Came Before the Big Bang?" *Washington Post*, January 4, 2016, https://www.washingtonpost.com/national/health-science/what-came-before-the-big-bang/2016/01/04/e8ca3606-ae52-11e5-b820-eea4d64be2a1_story.html?utm_term=.7ed9143e6cd8.

13  *A Kalahari Family*, dir. John Marshall (Watertown, MA: Documentary Educational Resources, 2002); E.M. Thomas, *The Old Way: A Story of the First People* (New York: Farrar, Straus, and Giroux, 2006).

14  Wilson, *Biophilia*; S.R. Kellert and E.O. Wilson, eds., *The Biophilia Hypothesis* (Washington, DC: Island Press, 1993); P.H. Kahn, Jr., "Developmental Psychology and the Biophilia Hypothesis: Children's Affiliation with Nature," *Developmental Review* 17 (1997): 1–61.

15  Wikipedia, s.v. "Radio Telescope," https://en.wikipedia.org/w/index.php?title=Radio_telescope&oldid=763519609 (accessed February 14, 2017).

16  National Radio Astronomy Observatory, "What Are Radio Telescopes?" https://public.nrao.edu/telescopes/radio-telescopes (accessed February 14, 2017).

17  P.R. Ehrlich and A.H. Ehrlich, *The Dominant Animal: Human Evolution and the Environment* (Washington, DC: Island Press, 2008); Kahn, *Technological Nature*.

18  P.H. Kahn, Jr. and P.H. Hasbach, "The Rewilding of the Human Species," in *The Rediscovery of the Wild*, ed. P.H. Kahn, Jr. and P.H. Hasbach, 207–32 (Cambridge, MA: MIT Press, 2013).

19  T. Hartig and P.H. Kahn, Jr., "Living in Cities, Naturally," *Science* 352 (2016): 938–40, doi:10.1126/science.aaf3759.

20  P.H. Kahn, Jr. et al., "Human-Nature Interaction Patterns: Constituents of a Nature Language for Environmental Sustainability," *Journal of Biourbanism* 17, nos. 1 and 2 (2018): 41–57; P.H. Kahn, Jr., J.H. Ruckert, and P.H. Hasbach, "A Nature Language," in *Ecopsychology: Science, Totems, and the Technological Species*, ed. P.H. Kahn, Jr. and P.H. Hasbach, 55–77 (Cambridge, MA: MIT Press, 2012); P.H. Kahn Jr., et al., "A Nature Language: An Agenda to Catalog, Save, and Recover Patterns of Human-Nature Interaction," *Ecopsychology* 2 (2010): 59–66; P.H. Kahn, Jr. and T. Weiss, "The Importance of Children Interacting with Big Nature," *Children, Youth, and Environments* 27, no. 2 (2017): 7–24; P.H. Kahn, Jr., T. Weiss, and K. Harrington, "Modeling Child-Nature Interaction in a Nature Preschool: A Proof of Concept," *Frontiers in Psychology* 9 (2018): 835, doi.org/10.3389/fpsyg.2018.00835.

21  S. Scheffler, *Consequentialism and Its Critics* (New York: Oxford University Press, 1988); B. Williams, *Ethics and the Limits of Philosophy* (Cambridge, MA: Harvard University Press, 1985).

*Part II*

# MORALITY *and* MORTALITY

# THREE

# A Ship from Delos

JAN ZWICKY

I N 399 BCE, SOCRATES, THE ATHENIAN PHILOSOPHER CONDEMNED
to death for crimes against the state, had to wait some time in prison before
being executed. Plato, in his dialogue *Phaedo*, describes the circumstances:

> *Phaedo*: ... As luck would have it, the day before the trial, the stern of the
> ship—the one the Athenians send to Delos—had been wreathed with
> garlands.

> *Echecrates*: What ship is this?

> *Phaedo*: It's the one—so say the Athenians—in which Theseus once went
> to Crete, taking the two lots of seven victims; and he brought them
> to safety and was himself saved. It is said that he vowed to Apollo at
> the time that if they were saved he would lead an embassy to Delos
> every year. Because of this, always, year after year, even now, they send
> this embassy to the god. It is their law that once the embassy begins,
> during its course the city is to be pure and no one can be executed until

such time as the ship shall have arrived at Delos and returned....The
embassy begins when the priest of Apollo has wreathed the stern of the
ship. This happened, as I say, the day before the trial started. (58a–c)

And here is Socrates' reaction to the news that the returning ship had been
sighted at Sounion:

Socrates [to his friend]:...Why have you come so early?

Crito: I bring news, Socrates, hard news. Not for you, apparently, but for
me and for all your friends, hard and grievous news, news which I
myself would take to be among the most grievous.

Socrates: What is it? Or has the ship arrived from Delos, the one on whose
return I must die?

Crito: It has not arrived yet, but I believe it will come today, from the
reports of some men who have come from Sounion and left it there. It
is clear from these messengers that it will come today, and so tomorrow,
Socrates, indeed it is necessary that your life must end.

Socrates: Yet, Crito, it may be for the best; if it is what the gods want, let it
be so. (Crito 43c–d)

Humans collectively are now in Socrates' position: the ship with the black
sails has been sighted. Catastrophic global ecological collapse is on the horizon. There is fine talk issuing from at least some political rostra. But nothing
remotely like adequate practical measures are being undertaken. In October
of 2016, Laurence C. Smith was still laying out the basic facts of climate change
for the educated readers of the New York Review of Books. "Many are surprised
to learn that the science of greenhouse warming is uncontroversial," he wrote
in a review of Tim Flannery's Atmosphere of Hope, a book that promotes "third
way" technologies as solutions to the crisis. On November 29, 2016, Canada's
prime minister, Justin Trudeau, approved two major pipelines that will allow
the Alberta tar sands to boost production. In the same week, The Economist—
which, though a radical defender of free markets, has been ahead of the curve

in its open-eyed acknowledgement of the impacts of climate change—projected a world-wide demand for oil over the next two decades that would raise temperatures by 5 degrees Celsius.[1] President Donald Trump has publicly stated that he believes climate change is a hoax, and originally placed Scott Pruitt, a climate-change denier, in charge of the Environmental Protection Agency. Pruitt has since been forced to resign amid scandals and has been replaced by Andrew Wheeler, a former coal lobbyist. Myron Ebell, who led Trump's EPA transition team, has said that Wheeler is "as fully committed to advancing [Trump's agenda] as Pruitt was."[2] In late December 2016, respected economist William Nordhaus published a discussion paper with updated projections on climate change, concluding that "it will be extremely difficult to [hold global warming to 2 degrees Celsius] even if ambitious policies are introduced in the near term."[3] Yet on June 1, 2017—the day oil began to flow in the Dakota Access Pipeline—Trump announced that the United States would withdraw from the 2015 Paris climate accord. On February 2, 2018, *Newsweek* reported that in April 2017 Pruitt personally oversaw the erasure of information on climate change from the EPA's website. A month earlier, the EU Climate Leader Board had reported that Europe, as a whole, was not on track to meet its Paris commitments.[4]

Populations of wild animals, including shellfish, fish, and pollinators, on whom humans depend for food, are crashing. Coastal communities are being inundated, their supplies of fresh water turning saline. Continental supplies of fresh water, previously stored in glaciers, are disappearing.[5]

Socrates, many of us believe, was innocent of the charges on which he stood condemned. I do not believe this is true of many citizens in rich industrialized countries, who live comfortable, air-conditioned lives, surrounded by a vast array of plastics and energy-consuming conveniences, who drive SUVs, travel frequently by air, have several children, and eat a lot of meat and imported food. But when those citizens go, myself included, they're going to take a lot of innocent beings with them.

What could constitute virtue in the circumstances in which we find ourselves? The answer is surprisingly straightforward: what has constituted virtue all along. We should approach the coming cataclysm as we ought to have approached life.

There are, of course, many moral traditions and many differences of emphasis among them. I am not an anthropologist and cannot offer a thorough comparative analysis. I propose, then, the suite of virtues that Socrates himself was said to have cultivated and which, according to Plato, he embodied so clearly on the day when he knew he was going to die. These virtues have the added interest of lying at the foundation of moral thought in the industrial, post-industrial, post-colonial consumer societies whose behaviour, arguably, has precipitated the crisis.

## Excellent Human Being

The English word *virtue*—from the Latin *virtūs*, meaning *manliness*—is frequently used to translate the Greek word ἀρετή (*aretē*). In English, *virtue*, through Christian influence, has come to mean almost the opposite of its Latin root: it connotes the stereotypically feminine traits of meekness, purity, and quiet obedience. The Greek concept, by contrast, emphasizes *excellence*, without explicit reference to sex. (Or, indeed, to species.) To possess ἀρετή was to excel at being the thing you were. It was to be a noble exemplar.

General human excellence for the Greeks was comprised of several specific excellences. In *Republic*, Book IV, Plato selects four of these for special attention: σοφία (*sophia*), σωφροσύνη (*sōphrosunē*), ἀνδρεία (*andreia*), and δικαιοσύνη (*dikaiosunē*). All four are very much on display in Socrates' final hours. (It is a short list by Greek standards, but it contains no suggestions a Greek would have disputed.) I am comfortable translating the last two as *manly spirit*—or, setting gender aside, *courage*—and *justice*; but the first two have no straightforward English cognates.

*Sophia* is usually translated as *wisdom*; but in Greek it means, first and foremost, skill in practical tasks. It is often equated with *phronēsis*, the savvy of the experienced man of affairs. To possess *sophia*, then, is to *know what's what* and to act accordingly. It is to *be aware*. Thus it is striking that Plato insists throughout his corpus that Socrates' *sophia* was rooted in his awareness that he did *not* know. This aspect of Socrates' character is best described not as ignorance but as *humility*—a virtue not much admired by male Greek culture, which may explain Plato's refusal to treat it explicitly as such. What Socrates displays is not humility in the sense of abasement or meekness, but a robust willingness to ask himself why he believes what he believes, and an equally robust refusal to regard "because everyone else believes it" as a good answer. Socrates did not let concern for his social reputation cramp his epistemological style. Although I think we can and should distinguish humility as a distinct ecological virtue, I also think we should follow Plato's lead in connecting it first and foremost to knowing what we do, and don't, know.

*Sōphrosunē*, the fourth cardinal Socratic virtue, is sometimes translated *temperance* or *moderation*, occasionally *soberness*. We can see what these translations are driving at, but they convey a stodginess, a primness, that is absent from the Greek meaning. In Greek, the word means *of sound mind*, and the dictionary notes that it is closely related to, and often interchangeable with, *emphrōn*, to be *in* one's right mind. I will translate it as *self-control*, where this is understood to have particular reference to physical indulgence.

In *Protagoras*, Plato adds a fifth virtue to his canonical list, ὁσιότης (*hosiotēs*). (He also substitutes the word *epistēmē*, which means simply *knowledge*, for *sophia*.) *Hosiotēs* is usually translated *piety*, but the connotations of rigidity or dourness that accompany the contemporary English term are no part of the Greek meaning. *Respect for divine law* would be closer, especially if one added that divine law is often inscrutable. *Phaedo*, the dialogue in which the ship from Delos has arrived, is framed by explicit spiritual observance on Socrates' part, so I think we must add some version of such observance to the list. In an attempt to create an ecumenical moniker, I will call it *contemplative practice*.

Finally, in one of the most moving passages in Plato's work, Socrates is shown to demonstrate an alert but impersonal compassion for his friends. Midway through *Phaedo*, Socrates' argument that the soul is immortal and that therefore no one should grieve his death appears to have fallen to two powerful objections. (Objections that Socrates himself has had to force out into the open, as the young men harbouring them have been afraid to voice them.)

> *Phaedo*: Indeed, Echecrates, I have often greatly admired Socrates, but
> never as much as on that occasion. It wasn't strange perhaps that he
> had something to say. But what I marvelled at most about him was how
> pleasantly and graciously and admiringly he received the young men's
> argument, and then how keenly he sensed its effect on us, and then how
> well he healed us and, as though having called us back from our fleeing
> and defeat, brought us around to contemplate the argument together
> with him. (88e–89a)

We arrive at the following list of core Socratic virtues:

- knowing what's what: awareness coupled with humility regarding what one knows
- courage
- self-control
- justice
- contemplative practice
- compassion

How might a human being living now or in the near future manifest these excellences, in the face of the sighting of our own global ecological ship from Delos?

## Knowing What's What

### *Awareness and Humility*

Looking the truth of our situation in the eye means understanding this:

> Being will be here.
> Beauty will be here.
> But this beauty that visits us now will be gone.[6]

Really to be aware is to hold this knowledge inside every thought, every gesture.

This is, of course, what Buddhist and Taoist traditions call enlightenment. And that's a tall order. Genuine enlightenment includes an acknowledgement that humans are fallible—even saints. Nonetheless, it is this limpid recognition of mortality that we are striving for. It is to look at the world openly and to see it, and one's own actions, and the actions of others, for what they are: gestures that vanish in the air like music. We keep our reactions to this recognition—especially the reaction of fear—balanced by exercising the other virtues.

The objection that is often raised to looking the facts in the eye is that doing so will result in despair; it will quash hope and thereby make our lives unbearable. Hope, it is argued, is one of the great virtues, too. And in these circumstances, it must be allowed to override knowledge, because knowledge is just too painful.

Hope *is* one of the great virtues. Socrates himself manifests it in his abjuration to his friends after he is sentenced at his trial: he says that there is "much hope" that death itself is good (*Apology* 40c). But to imagine that awareness of the facts precludes hope is to misconstrue both awareness and hope. To be aware that death is imminent is not to wallow in despair; it is precisely not that. To be aware is to acknowledge what is the case. What is the case for many humans—those with enough to eat and a roof over their heads—includes great beauty: the natural world, works of the human spirit and imagination, the love we feel for others and the love they feel for us. To wallow in despair that the natural world is dying is to fail to be aware that it is still, in many ways, very much alive. It is also to fail to understand that in precipitating drastic climate change and a sixth mass extinction, industrialized humans are not destroying everything. Being will be here. Beauty will be here.

One form that hope takes in this situation is knowledge that the earth is prodigious. After other mass extinctions, life has proliferated again. After civilizations have collapsed, others have arisen. Perhaps there will be some new species with many of our talents and fewer of our vices.

Hope also, in all cases, takes the form of humility. We are not, for the most part, aware of the future. It is not usually perceptible in the way that this sunset, this windiness in the trees, the touch of this hand on our own, is perceptible. Nor is it fully predictable on the basis of the knowledge we have: the future as a whole is like the weather. Most importantly, the resonant and non-linear connections within being, those recognized in numerous Asian and Indigenous traditions, are not graspable by the calculative intellect, which undergirds technocracy. To hope for a techno-fix is to imagine, yet again, that calculative rationality can control the world; it is hubris. Humility means recognizing, clearly, that we don't understand everything. We do not know what joys may accompany us, nor what possibilities the situation contains.

What of humans who do not have enough to eat, or those who, like females in many cultures, suffer systematic injustice, or those, female and male, whose cultures have been systematically savaged by colonial empires? In what do awareness, humility, and hope consist for them? It is hard to say that they, too, must reach for enlightenment when they are simply trying to stay alive. I will not say it. But then those of us who do have enough to eat and the time and freedom to think must see clearly what their situation is. Its desperate character, its blinding pain, must become an integral part of what we know: that humans undergo this at the hands of other humans; that they suffer this and force this suffering on others. And that we, who are this moment reading this volume, do not know everything. And that beauty exists.

What about remorse? The Buddhist tradition teaches that change is constant: *this beauty that visits us now will be gone.* But doesn't it make a difference if one's own deliberate and self-regarding actions have wrecked things? It does. Denial of responsibility, denial of complicity is part of refusing to know what's what. It blocks the kind of grief that can realign the soul, that allows us, ultimately, to be free.

Indeed, accepting responsibility—wanting to co-respond—*just is* to understand. Simone Weil writes: "The poet produces the beautiful by fixing attention on something real. The same with the act of love. To know that that man, who is hungry and thirsty, truly exists as much as I do—that is enough, the rest follows of itself."[7] Love, awareness, and the desire to respond: these are distinguishable but inseparable aspects of genuine intelligence.

And if what one has ruined is ruined beyond repair? If restitution is not possible? What a difficult case! Compassion for the person who finds herself in the clear light of such knowledge. That knowledge will strike deep into the body as sorrow and remorse. Again, exercise of other virtues is crucial if something positive is to come of such grief. But awareness itself also prevents paralysis. What use, to anyone, is it to lie down, immobilized by pain? Pain must be used to turn the soul toward the real, to reform both action and attention: to love what, in this case, remains.

## Courage

The role of courage in the face of cataclysmic environmental change is obvious. It will take physical courage to stand up to the inevitable physical pain and duress. It will take civic courage to live with the wars and civil wars, the death tolls and disease, the mob violence that are likely to become common. But above all, it will require moral courage to continue to exercise the virtue of awareness: to see things for what they are, and to know, *when* one does not know, *that* one does not know: to refuse to conflate opinion and ideology with understanding.

Humility—a deep unconcern with the social fate of the self—is the foundation of courage as well as wisdom: it frees one to see the truth. Humility and courage empty the mind of irrelevant preoccupations and clear a space for insight.

## Self-Control

Self-control appears to be a virtue that humans—at least those who construe themselves as 'consumers'—severely lack. Its fundamental importance is taught over and over in the stories told by sustainable cultures, which suggests that it may be difficult for any human being to acquire. *Pleonexia*—wanting more than enough—is Plato's name for the complementary vice, which appears to be the defining characteristic of North American and other post-colonial cultures.

Self-control is behind reducing, reusing, and recycling. It's behind wearing old clothes, letting those clothes air-dry, and wearing a sweater instead of turning up the thermostat. It's behind eating locally, eating in season, and staying away from sugar and fast food. Self-control can prompt one to walk or take the bus, to choose to have no or very few children, and to vacation at home or very near by. Altogether, it allows a joyous simplicity, a delight in living as lightly as possible on the earth. One gives away what one does not need.

The joy and delight that attend genuine self-control show that there's an intimate connection with awareness. Grim, Procrustean self-denial, self-excoriation with rules one has adopted without examining—this is ideological repression, not self-control. Awareness reforms desire; or, rather, it allows other desires—for the well-being of others, non-human and human—to become immediate and powerful. *To know that that man, who is hungry and thirsty, truly exists as much as I do—that is enough, the rest follows of itself.* Self-control is, in a way, not control at all: it is the melting away of the need for certain forms of comfort and distraction. It is an *embrace* of simplicity. When we know what's what, pleonexia doesn't arise. Our understanding of happiness alters. We actively desire the health of the ecological community to which we belong. We want to do what it takes to be at home.

## Justice

Plato's *Republic* is about justice. It is an attempt to develop an account of both the just state and the just individual that will allow Plato to defeat the skeptical challenge with which the dialogue opens: Why be just if you can get away with murder (and graft and embezzlement and adultery)? Why be just if you're going to end up attacked by the mob anyway? Plato's response locates justice in the harmonious relations among the parts of the tripartite soul, each of which mirrors a distinct class in the ideally efficient state. It's not an account of justice that speaks to intuitions about punishments fitting crimes or equitable distributions of goods. I won't try to defend it here. I mention it, though, because it allows us to understand how, in Plato's conception, the four cardinal virtues are facets of an integrated whole. *Sophia*, or knowing what's what, is the core virtue of the mind—or the intellectual class, which Plato puts in charge of running the state. *Courage*, unsurprisingly, is the core virtue of the 'spirited' or warrior class, which is in charge of the state's defence. *Self-control* is the virtue of the 'appetitive' part of the soul, which corresponds to the class responsible for keeping the state ticking over physically: the farmers, cobblers, tanners, weavers, carpenters, and stone masons. *Justice* is manifest in the whole soul or state, Plato argues, when each part submits willingly to the direction of the intellectual faculty or class: "to produce justice is to establish an order in the soul in which parts rule over one another and are ruled according to nature; and injustice is those parts governing and being governed contrary to nature" (444d). It is only armed with this conception of justice that Plato is able to defeat the initial challenge and to argue that the just person, even if wrongly perceived as unjust (read: Socrates), will have an objectively better life than the unjust person who succeeds in getting away with heinous crimes. The life of the unjust person can't be worth living, Plato claims, when his soul—the very principle by which he lives—"is corrupted and in turmoil" (445b).

This is not how we usually think of justice. We usually think of it as John Rawls did, as fairness. Justice requires a fair distribution of goods, we say, a fair legal procedure, a fair apportioning of work, requisite punishments for transgressions. Rawls suggested that we can find out what sorts of arrangements are fair by asking people to choose the situations in which they want to live behind a so-called 'veil of ignorance'—the veil allows them to know everything about various hypothetical social, political, and economic arrangements except what role or position they themselves will occupy. No one who's rational is going to vote for an arrangement in which there's a good chance they'll end up as a slave, or a disenfranchised Aboriginal, or a female non-person. Those kinds of social arrangements, we conclude, are manifestly unjust since no one not already born into them would risk choosing them.

How might either Plato's notion of justice as interior harmony, or the more familiar notion of justice as fairness, govern our behaviour in the context of cataclysmic ecological change?

Justice as interior harmony effectively summarizes the internal relations we've already noticed among awareness, humility, courage, and self-control. Humility—getting the ego and its fears out of the way—gives one the courage to seek truth; it helps one discern where one must press further. Awareness makes self-control easy: it turns it from an onerous task into a series of self-reinforcing behaviours that allow one to feel at home. The resulting simplicity supports humility; awareness widens and courage builds. It was Plato's intuition that this interdependence, which fosters the full expression of each virtue, was the secret of Socrates' moral charisma: it was the organic interacting complex of specific virtues that made him the most excellent of men. The complex was manifestly stable and self-reinforcing. It allowed Socrates to suffer injustice, physical pain, imprisonment, and death with sprightly equanimity. If 'justice' seems the wrong name for this virtue, call it something else: nobility; integrity; shiningness. What produces it is the self-sustaining interdependence of awareness, humility, courage, and self-control.

Justice in the sense of fairness—equal access to the necessities of life, equality before the law—has obvious application to our present and pending situations. It figures notably in attempts to get fully industrialized nations to register that the effects of climate change are already being borne more immediately and more severely by human populations that have done the least to cause them. Will the governments of industrialized nations act generously? The global rise of nationalist protectionism suggests not.

Can individuals in rich nations do anything besides pressure their governments to act?

They can live simply, with less, and offer more to charity. They can divest themselves of investments in fossil fuels, neonicotinoid pesticides, and clear-cuts. They can have few or no children. They can sponsor climate refugees. They can ask themselves what, were positions reversed, they would want rich individuals to offer them. And then carry through.

Such actions will not be enough, though. They will not be nearly enough. And they speak only obliquely to the devastation of non-human lives that underlies the pressure on human ones.

This raises a deep question: Since we cannot remedy the situation, why bother with justice at all? Here is where Plato's alternative conception of justice has real bite. It is an attempt to respond to a common intuition that moral gestures matter even if they do not have a noticeable impact on the state of the world. About justice specifically, Plato argues that it is a 'right ordering' of the soul, independent of circumstance; he claims that it will be experienced as such by the person whose soul is so ordered and will be perceptible as such by others. It is, in this sense, its own reward. The bedrock under this claim is Plato's conviction that the good is beautiful, that it is what most deeply and unquestioningly draws the soul. While the metaphysics of this view have been disputed for centuries, and while the view itself is presently in relativist eclipse, one of the things that's interesting about it is that it doesn't go away. Another thing that's interesting about it is that it's rooted, in Plato's account, in descriptions of how it *feels* to encounter the good: it is a kind of falling in love, mysterious, powerful, and life-changing.[8] These descriptions strike many people, even now, as accurate.

At the same time, there can be no denying that the history of human sadism, cruelty, violence, exploitation, and uncaring is as old as the species. How could anyone, in the face of that history, claim that humans naturally and immediately love the good?—By arguing that when they appear to disregard the good, they can't *see* it. The soul's eye, Plato claimed, is often buried in a muck of ignorance and miseducation.[9] But when that eye is clear, even in the most appalling situations, people respond with alacrity to perception of the good. *To know that that man, who is hungry and thirsty, truly exists as much as I do—that is enough, the rest follows of itself.*

"Very convenient," an interlocutor might respond, "anything that looks like counter-evidence gets waved away as the effect of 'ignorance.' " I'm sympathetic to this objection. But I also have to report that the older I get, the more it looks like Plato was on to something: there are people in all walks of life who, otherwise inexplicably, are seized by the need to do the right thing. Nietzsche and other advocates of postmodern cynicism argue that the appearance of 'good' behaviour is simply the result of kowtowing to social rules erected by those in power to serve their own ends. But when you're in the presence of someone who is acting from direct perception of the good, you can tell; and you can tell that Nietzsche's explanation is inadequate. There's nothing rote, or cowed, or obsequious about it. Such action is overwhelmingly—breathtakingly, beautifully—*free*.

## Compassion

Compassion, like humility, does not usually loom large in Greek accounts of human excellence. There is no question, however, that Socrates feels it for the men who will witness his death and who must then carry on living without him.

In our own context, as in Socrates' prison cell, it is extremely important to offer compassion to those struggling to come to awareness. Contempt for fear is not only graceless, it is damaging: it intensifies anxiety, thereby intensifying denial.

It is also important to distinguish between compassion and displays of affection. In some circumstances, compassion may involve or even require overt kindness or concern. But as Socrates understood, in other circumstances, emotional display may be disabling. The goal of compassion is the alleviation of suffering. This is sometimes achieved by offering companionship in suffering. Often, though, what helps the most is empowerment, disinterested assistance toward clear understanding, courage, and self-control.

## Contemplative Practice

In more than one dialogue, Plato shows Socrates to be well-informed about Greek myth and religious ritual. (In *Phaedrus*, he is shown to know something about Egyptian myth as well.) He pays attention to his dreams and tries to do as they command; he tries to obey the Delphic oracle; he is genuine and generous in his ritual observance, making appropriate prayers to appropriate deities. And then there is his sign or *daimōn*. It visits him, he claims, whenever he is about to do something that is not right.

In the references that I have found, this *daimōn* is not usually represented as a voice but merely as something that forbids action.[10] Paul Shorey describes it "as a kind of spiritual tact[,] checking Socrates from any act opposed to his true moral and intellectual interests."[11] At any rate, Socrates claims that its origin is divine, and he always obeys it. He associates it with his prophetic abilities, though he adds that these are modest.

At the heart of contemplative practice of any sort is attention. As Weil observes, prayer is nothing other than absolutely unmixed attention.[12] Attention to anything? To the stock market? To social media? Unmixed attention to human institutions and fashions is perhaps a kind of prayer, but it is an abject one: the plea of the infant ego. In contemplative practice, the ego does not play a big role. Instead, we attend to the real, physical world, its immense and intricate workings, its subtlety; its power, its harshness, and its enormous beauty. We attend to the miracle of it, that there is something—this, here, now—rather than nothing. We attend to the rhythms of this world as they play out in our lives and the lives of those we know: pain, sex, birth, death, love. We slow down enough that we can sense these rhythms.

We attend also to the world's extraordinary surprise: its refusal to quit, the weed flowering in tar, the way beauty and brokenness so often go together. If we are among the lucky with enough to eat, genuine companionship, a dry bed, it is easy to make ourselves available to these lessons; we are shaped, spontaneously, by gratitude. As we become more adept at contemplation, we find we can perceive beauty even when we are hungry or tired or lonely. Loneliness is transmuted to solitude, and then to a sense of companionship with non-human being; contemplative practice deepens again. Gratitude deepens. The imagination, and with it our availability to what is real, expands.

The more we attend to the world, the less we find ourselves wishing to control it. In place of that wish, a different wish grows: to become, as Aldo Leopold puts it, a "plain member and citizen" of the land-community.[13] Wonder grows, too, which breeds respect. It also breeds a willingness to take intuitive forms of knowing seriously. We start to become attuned to the world's resonance. I have not met a contemplative who does not experience this attunement as love.

Does contemplative practice issue in religious commitment? Sometimes, but by no means always. Does it demand ritual? No, though often spontaneous ritual observances arise around gratitude, grief, and expressions of love.

Members of technocratic cultures worldwide have a lot of mourning to do. "Mourning draws on transcendent but representable justice," says Gillian Rose, "which makes the suffering of immediate experience visible and speakable."[14] It returns the soul to its community. A few large, collective expressions of remorse might not be out of place at this point. But the deep acknowledgement will come in private meditation and conversation, when we point with our hearts to what we have destroyed, to our addictions and to our self-deception about our addictions. Only that acknowledgement, that prayer, can free us into real and cleansing grief.

## Why Don't We Do What We Know We Ought To?

We don't, most of us, possess these virtues to the degree that Socrates did. It isn't news that we ought to try to acquire them, though. And the fact that over the course of millennia, imperial cultures in Europe and Asia have repeatedly failed to inculcate these (or similar) virtues in most of their powerful citizens is sobering. It suggests that, as a species, we're not very good at taking our own advice. Why is this?

I don't know. It's an old, old puzzle, weakness of the will, or *akrasia*, as the Greeks called it. Plato's view is that we don't do what we ought to because we misjudge the facts of the case.[15] In particular, we judge, incorrectly, that the pleasure that attends some immediate gratification (jetting to Mexico for a holiday) outweighs the pain that attends some future scenario (global warming). It is a version of the view that ignorance is the root of evil. We don't, in fact, *see*—with the rehabilitated eye of our own souls—that we should keep our feet on the ground at home. We mouth the moral injunctions fed to us by our society, but we have not *discovered* them as life-determining truths. As Plato would agree, however, this response is ultimately circular, for genuinely knowing what's what and having the courage to admit it are crucial aspects of virtue. To say that we fail in virtue because we fail in virtue is not an explanation.

Plato also says that the question of why we fail to do what we know we ought to is one on which all other insights about the nature of virtue depend.[16] For virtue concerns excellent human *being*—here, now, in this world. There is little point asking how to live or how to die if we won't, or can't, act on the answer. We know that awareness of the good does genuinely compel many of us some of the time. And it appears to compel a few of us, like Socrates, all of the time. But it doesn't seem to compel most of us most of the time. This may simply be a brute fact about the species.

At the core of the puzzle is the apparent failure of knowledge to lead, at least on some occasions, to the sort of action that Weil so eloquently describes. We recognize what she means when she says that registering someone else's pain is "enough"—that action to remedy that pain follows instinctively. Most of us have had this experience. And yet we also know that sometimes, that *often*, such knowing *isn't* enough. Why didn't we citizens of rich nations impose mindful constraints on our own consumption when the science came in decades ago? As I recall, people claimed that both cycling and recycling were inconvenient;

and they didn't want to jeopardize careers or social esteem by refusing to fly or to wear fashionable clothes; they wanted to make a lot of money, so they invested in Big Oil. It appears, in retrospect, that we were selfish, and lazy, and afraid of social censure. We knew, we knew well enough to be made uncomfortable by our knowledge, but we didn't *want* to know. Our awareness was weightless. So weightless perhaps that if we behaved as though it didn't exist, it would go away.

We see once again that there is no sharp distinction between awareness and justice conceived as integrity; it routinely takes courage and self-control—steadiness of vision in the face of fear or shock or disbelief—to admit what we know, just as it takes courage and humility to admit what we don't know. Part of what Weil's observation tells us is that knowledge, if it is to be morally excellent, must be taken into the heart as well as into the mind. Or, to put this another way, that for the virtues to *be* virtues, they must be practised in concert. Acting from one necessarily calls others into play. Becoming an excellent human being requires one to adopt a moral ecology; and ecologies aren't modular constructions. They don't consist of separate bricks that are snapped together (or unsnapped) like pieces of Lego. Moral ecologies, like biological ones, are organic wholes whose distinguishable aspects—the virtues—stand in internal relations to one another.

None of awareness, courage, humility, self-control, compassion, justice, or contemplative practice can be acquired without acquiring the others.

## Epilogue

How are we to die? This question is at the heart of Plato's portrait of Socrates. More even than the dialogic form in which he wrote, its presence points to a profound connection between Plato's philosophical genius and the great tradition of Athenian tragedy.[17]

Lists from the Athenian Dionysia indicate that tragedians usually presented their plays in quartets: three linked tragedies plus a satyr play—a short, rude, and rustic piece whose purpose seems to have been to provide comic relief amid the doom and gloom.

A sense of humour does not appear as such in the ecology of Socratic virtues that we've been considering. Socrates, affable and ironic by turns, never tells a thigh-slapper; nor did Plato compose short, self-contained philosophic romps. But a sense of humour needn't include a propensity for telling ribald jokes; and I propose that we follow the tragedians' lead, at least in principle. Given the other virtues with which it must be coordinated, a Socratic sense of humour will be manifest not as slapstick and belly laughs, but as the lightness of touch that comes from not taking one's self too seriously. We will sense it as a smile: the absence of fear and the refusal of despair. Even in the face of death.

Like that pilot whose plane crashed. It doesn't matter if the story is apocryphal. After the crash, investigators were listening to the recovered flight recorder, and instead of the usual terror and yelling, they discovered that the pilot's last words were engagingly clear and calm.

"Point your toes, Herb," you can hear him saying. "We're going in deep."

## Acknowledgements

An earlier version of this essay was published by University of Regina Press in 2018 as part of a volume titled *Learning to Die: Wisdom in the Age of Climate Crisis*, co-authored with Robert Bringhurst.

## Notes

1   See graph entitled "The Big Picture," in *The Economist: Special Report: Oil*, November 26–December 2, 2016, p. 4.

2   William Nordhaus, "Projections and Uncertainties about Climate Change in an Era of Minimal Climate Policies," Cowles Foundation Discussion Paper No. 2057, December 2, 2016.

3   Oliver Milman, "Andrew Wheeler: 'Point man for Trump' focused on undoing Obama's EPA agenda," *The Guardian*, July 6, 2018.

4   See https://carbonmarketwatch.org/wp-content/uploads/2017/03/EU-Climate-Leader-Board_infographic_final.pdf. See also http://www.iflscience.com/environment/only-three-eu-countries-on-track-to-meet-paris-climate-agreement-targets/.

5   On shellfish and fish populations: a quick overview can be found in Craig Welch, "Ocean acidification, the lesser-known twin of climate change, threatens to scramble marine life on a scale almost too big to fathom," *Seattle Times*,

September 12, 2013. For scholarly treatments, see, for example, Benjamin S. Halpern et al., "Evaluating and Ranking the Vulnerability of Global Marine Ecosystems to Anthropogenic Threats," *Conservation Biology* 21, no. 5 (2007): 1301–15; Robert H. Byrne et al., "Direct observations of basin-wide acidification of the North Pacific Ocean," *Geophysical Research Letters* 37, no. 2 (2010): L02601, doi:10.1029/2009GL040999; and Nina Bednaršek et al., "New ocean, new needs: Application of pteropod shell dissolution as a biological indicator for marine resource management," *Ecological Indicators* 76 (2017): 240–4.

On pollinator populations: see Caspar A. Hallmann et al., "More than 75 percent decline over 27 years in total flying insect biomass in protected areas," PLOS *One*, October 18, 2017, doi:10.1371/journal.pone.0185809.

On coastal community inundation: see, for example, "Sinking shoreline threatens coastal communities in Indonesia," Reuters, March 6, 2018, https://www.reuters.com/article/us-indonesia-coastline-widerimage/sinking-shoreline-threatens-coastal-communities-in-indonesia-idUSKCN1GJ03M; and Erika Spanger-Seigfried et al., "When Rising Seas Hit Home: Hard Choices Ahead for Hundreds of US Coastal Communities," Union of Concerned Scientists, July 2017, https://www.ucsusa.org/global-warming/global-warming-impacts/when-rising-seas-hit-home-chronic-inundation-from-sea-level-rise.

On disappearing glaciers: see Twila Moon, "Saying goodbye to Glaciers," *Science* 356, no. 6338 (2017): 580–1, doi:10.1126/science.aam9625.

6  Robert Bringhurst, "New World Suite N⁰ 3," *Selected Poems* (Kentville, NS: Gaspereau, 2009).

7  Simone Weil, *La pesanteur et la grâce*, avec une préface par Gustave Thibon (Paris: Librairie Plon, 1988 [1947]), 196; available in English as *Gravity and Grace*, ed. Gustave Thibon, trans. Emma Craufurd (London: Routledge, 1987 [1952]), 108. There are many editions; this quotation comes from the section entitled "L'Attention et la volonté" (Attention and Will).

8  Plato, *Symposium* 201d–212c and 215b–222b, and *Phaedrus* 246d–256d.

9  This is the core of Plato's moral and epistemological vision, and is touched on, sometimes directly, sometimes indirectly, in nearly everything he wrote. Dialogues in which it is especially explicit include *Meno*, *Phaedrus*, and *Republic*. For the image of the soul's eye buried in the muck of ignorance, see *Republic* 533d.

10  *Alkibiades* 103a, *Euthydemus* 272e, *Euthyphro* 3b, *Phaedrus* 242b–c, *Republic* 496c, *Theatetus* 151a. See also Xenophon: *Defence* 12–13; *Memorabilia* 1.1.2–5, 4.3.12–13, 4.8.1, 4.8.5–6, 4.8.11; *Symposium* 8.5.

11  Paul Shorey, ed., *Republic*, Books VI–X (Loeb edition, 1935): 52, note a.

12  Weil, *La pesanteur et la grâce*, 192 (French), 106 (English). This quotation is in the same section as the previous quotation (see note 7).

13 Aldo Leopold, "The Land Ethic," in *A Sand County Almanac* (New York: Ballantine/Oxford University Press, 1966), 240. Again, there are many editions; this quotation comes from the section entitled "The Community Concept."

14 Gillian Rose, "Athens and Jerusalem: A Tale of Three Cities," in *Mourning Becomes the Law: Philosophy and Representation* (Cambridge: Cambridge University Press, 1996), 36.

15 Plato, *Protagoras* 352d–361d.

16 Plato, *Protagoras* 354e.

17 For a detailed discussion, see Martha Nussbaum, *The Fragility of Goodness: Luck and Ethics in Greek Tragedy and Philosophy* (Cambridge: Cambridge University Press, 1986). Diogenes Laertius in Book III of *Lives of the Eminent Philosophers* tells a story in which he says that Plato was about to compete as a tragedian at the Dionysia, but after listening to Socrates he burnt his work and became Socrates' pupil.

# FOUR

# Thanatopsis:
# Death Literacy for the Living

DAVID GREENWOOD & MARGARET MCKEE

PRELUDE
## "And as to you corpse I think you are good manure"[1]

DEATH IS RARELY EMBRACED AS A CENTRAL THEME FOR ENVI-ronmental learning. The ecological sciences, and environmental studies generally, celebrate life, not death. Death mainly enters the ecological conversation as something coldly statistical, as in life cycle or population studies, or as something to be avoided, as in the death of species, ecosystems, or the planet itself. How is it that death, which is so ineluctably central to life, can be so absent from our ecological attention?

Today, the medical sciences promise "advances" that will extend the average human lifetime well past a hundred years. Some researchers predict that within the next century, new technologies and discoveries around the human genome could extend individual human lives indefinitely. Of course, such life-extending technologies are expensive and will be available only to an elite minority that can afford boutique medical care. The fantasy of living forever thus comes with the dark side of reinforcing both consumer individualism and social inequality.

While the search for immortality is an old theme in human cultures and mythologies, the denial of and disregard for death in contemporary society is a relatively new phenomenon. As today's anthropocentric science seeks victory over death, many people are shielded from death's presence by its social banishment to nursing homes and hospitals, where life may be extended but not necessarily enhanced. For others living amidst violence and poverty, life may be surrounded by death from social rather than natural causes. Death, then, becomes a constant reminder of human injustice and suffering, of the failure of life to develop and thrive in struggling human communities. Both of these distortions—the denial of death or its gratuitous ubiquity—relegate death to the shadows of our collective consciousness. And while death is a topic we may organize our lives around avoiding, the media is saturated with its image in the form of daily terror and catastrophe.

Contemporary mass culture can be described as simultaneously death-obsessed and death-averse, yet art and literature have always embraced death as a central theme of life: "To be, or not to be—that is the question." The great poets of life are also often the greatest poets of death: seekers bold enough to open toward the reality of death as a generative force rather than shrink from it in fear of the unknown. In his "Song of Myself," Walt Whitman provides a salutary example of what it might mean even to befriend death (see Aspiz[2] for a comprehensive analysis of Whitman's poetry of death):

> And as to you death, and you bitter hug of mortality…it is idle to try to
>     alarm me.…
> And as to you corpse I think you are good manure, but that does not offend me,
> I smell the white roses, sweet scented and growing,
> I reach to the leafy lips.…I reach to the polish'd breasts of melons.
>
> And as to you life, I reckon you are the leavings of many deaths,
> No doubt I have died myself ten thousand times before.[3]

This turn toward death and its mysteries we consider a necessary ecological virtue in a culture that is at once both obsessed with youth and hell-bent on ecological destruction.

1.

# Death Literacy Confronts the Disappearance of Death

We need a new story about life on this planet, and that story has to include the fact that we, and all living things, must die. How are we to understand life, its brevity, frailty, and impermanence, without literacy about death? How are we even to notice the harm we are causing the planet when we cannot bring ourselves to acknowledge death? Death of plant and animal species, death of ocean and forest ecosystems, death of human languages and cultures: so much death hides in plain sight, and somehow we've convinced ourselves to look away, to not see it. Death can teach us, if we pay attention, how to live with greater joy and gratitude for life. Death can embolden our resistance to the consumerism and materialism of our times and strengthen our resolve to protect the planet we love. How, then, did we lose the will to notice death?

Our collective stories tend to deny, distort, and sensationalize death. We are bombarded by death imagery in ways we've never had to endure before as a species—gruesome imagery in film and print, aimed at priming us for fear and disgust. On the one hand we hide death away and we're implicitly told not to look, but on the other hand we can't get enough of a kind of voyeuristic looking that does nothing to satisfy our deepest need to know and understand death, and everything to render us overwhelmed, numbed, and alienated from it.

For many children, enchantment about life and curiosity about death go hand in hand. Enchantment is our birthright: an ancestral gift coded in our DNA as a score for joyful attention to the world around us and harmonious living in the larger family of being. Children instinctively imagine themselves into the bodies and lives of other animals—crawling, flying, roaring their delight as they play.[4] They soar like clouds, sway like trees, stand tall like mountains. Naturally curious about what it is to be alive and how their lives are entwined with the lives of all living things, they are also curious about death. Their early experiences of death, which often involve the death of a pet, or an insect in the garden, or a plant or tree, are formative moments in their growing understanding. But just as the basic alphabet for comprehending the question of death is beginning to form in children, and their curiosity leads to tough questions and the need to know, modern adults become strangely silent. Death just disappears from conversation, or is talked about in hushed and euphemistic ways that only reinforce the vanishing act. Death has become rare in the lives of children and young people, and when it does happen, it is so medicalized that children are automatically alienated and frightened, or so cartoon-like, as in film and video games, that its full importance is obscured. Death is the quintessential "open secret." As Leonard Cohen sings in "Everybody Knows," we tell ourselves the lie that we will live forever to avoid the truth, thinking that the lie will spare us pain and suffering.

Advances in medicine have radically changed how, when, and where we die in North America. For most of human history, death was commonplace. People used to die much younger of conditions we can now prevent through immunization, sanitation, and diet, or can easily treat with drugs and surgery. The average life expectancy in the 1980s in North America was in the mid-seventies; by 2010 it rose to more than eighty, and it is still rising. While we now expect to live decades longer, there is a real possibility that our last years will be lived with dementia and/or chronic illness that require support and complex medical care. Many people who generations ago would have been cared for at home as they aged, by family members, neighbours, and friends, now spend their last months or years in nursing homes, where they die under the care of teams of professionals, often out of reach of the people who knew them all their lives. We are so estranged from the naturalness of death that an older person whose death is anticipated from a long-standing medical condition is likely to die after being admitted to the intensive care unit of an acute care hospital, rather than at home in the arms of their loved ones. Our collective imagination is so saturated with expectations of a medical fix for every disorder that we can't imagine a death that isn't preceded by some kind of *battle* or desperate search for *one more thing to try*.

Both ways of dying—in a nursing home after a slow decline, or in intensive care after what the obituaries refer to as "brave battles"—often leave family members bewildered and traumatized, and no closer to understanding or coming to terms with their own inevitable death. And sadly, children and young people are often prevented from being present at these deaths by parents who fear that exposure to death will traumatize them. The end result is that many children grow up without ever directly witnessing death. And the minimal instruction they do get is often drenched in euphemism and mystification, which do nothing to satisfy their need to understand death.

Change is coming. Palliative care is transforming medical care for people with life-limiting illness. There is a growing body of autobiographical/

testimonial writing from people who have documented their approaching deaths,[5] and a growing children's literature about death and dying. Grassroots efforts like the Compassionate Communities movement, Dying Matters, Groundswell, and other educational initiatives (including initiatives to train volunteers to work with the dying, and growing numbers of university and college courses on death) have done much to improve the quality of life for people with chronic, life-threatening illness, and the quality of the care they receive at the end of life.[6]

All of these efforts to teach about death have kick-started much-needed conversations in health care, and in families, schools, and whole communities. More children are learning about aging and death. More adults are reading, thinking, and talking about how and where they want to die and making explicit plans for their end of life. We have the beginning of new stories about how to die and how to take care of dying people. More people now arrive at the end of life knowing what to expect, with the kind of care they desire, and with the prospect of a good death.

If we could see death more clearly and understand its absolute necessity in our own lives, might we live more fully? Might we object more strenuously to every discourse and practice that hides aging and death? Might we just say "stop" to marketing practices that prey on our insecurities, exploit our desires, and promise us eternal youth? A full-bodied literacy about death will acknowledge our collective complicity in its disappearance. Death didn't just disappear. We ourselves performed the sleight of hand, and we alone can bring it back.

2.

## Death Literacy Teaches Emotional Literacy

Our collective silence about death is not the fertile kind of silence that comes from precious solitude; rather, it is the kind of culturally policed silence that makes certain experiences unspeakable, isolating and pathologizing those who do speak. This kind of silence does not protect us from pain; it worsens it. It does not spare our children "unnecessary" suffering. It leaves them unequipped, without a basic vocabulary, so that at the very times in life when they most desperately need to make meaning, and most long to seek comfort in connection, they will find themselves alone with pain that deepens in the silence.

The purpose of death literacy is not to tame death or to alleviate the pain of grief and loss. (That would actually be the purpose of *illiteracy*—to sanitize awareness and conversation of all that causes pain, to keep us in the dark about what we don't know.) If anything, death literacy will awaken the pain of death, and of all loss and grief, and dignify it with fuller awareness. To name this pain, to speak it and share it, is to restore it to its proper place in communal life, where it can be held and grieved in common. Literacy about death is a bold act of defiance—a way of turning toward pain (rather than away from it), and of allowing the reality of our vulnerability and the inevitability of loss and grief to burn through our emotional bodies. Talking about death is an act of courage, and listening an act of love: It is saying, "We are in this together. You are not alone."

Our emotions carry an embodied, ancestral wisdom for living with wonder, joy, and curiosity. It is not possible to selectively shut down the feelings we don't like (sorrow, anger, despair, fear) without also diminishing our passionate engagement with the rest of life. Anyone who has attended a wake knows that sorrow, laughter, and gratitude can fill a room at the same time, and that they flow from the same source.

The emotional life of children, when they are still in their bodies and before they are taught to suppress their feelings, is vibrant and alive. The challenge as we grow up is to remain in our bodies, unguarded in our vulnerability. It is not emotion per se that makes us suffer, but the management of it as a problem, and the consequent tendency to go around it, distract from it, or numb it. In a culture that so fears and denies powerful emotions and the bodily-felt chaos they bring, it's hardly surprising that powerful emotions, especially anger, fear, and grief, feel scary and unmanageable to many people. They propel us out of the orbit of friends and family, who may also be under the impression that it is better "not to talk too much" about these things, better to "handle them on your own." Not talking about emotions only adds to the suffering people feel when in the grip of powerful emotions.

This is especially true in the management of grief. For an experience so ubiquitous, one might expect more acceptance of its wildness and resistance to taming, and yet grief is one of the most policed of human emotional experiences. While there is no question that some grieving people are enormously helped by professional intervention (particularly if it is a relationship that offers a non-judgmental acceptance of grief's roller coaster), the rampant pathologizing of grief in North America, and the industry designed to "treat" it, has probably done more harm than good. Increasing numbers of people have been told that they are "doing it wrong," that their grief is too wild, and that they need to control it or medicate it. Increasing numbers of people are told they need professional intervention in order to learn how to do grief right in order to "recover." The idea that grief is a temporary problem from which one can (and should) recover is a profoundly damaging discourse that delegitimizes the lived experience of every person who has ever grieved. It heaps upon their already broken hearts the added burden of having to protect others from the chaos of their pain, and the added isolation that comes with their suffering being unspeakably difficult to express. What grieving people actually need is unlimited time and social support, and connection to a community where the chaos is welcomed and where there are rituals to give grief meaning and value. Instead, we ask for the chaos to be controlled and tidied up before it reminds us too painfully of our own vulnerability.

Death is a community event, not merely an individual and medical event. We aren't meant to do it alone. It takes a whole community and a robust emotional literacy to hold our sorrow, and to face our fears, together. If we do not learn to grieve, it becomes difficult to imagine how we will ever confront our collective ecocide. Emotional illiteracy guarantees that we will continue to look away.

3.

# Death Literacy Teaches Biological and Evolutionary Thinking, Linking Humans to Other Living Beings in the Larger Story of Life on Earth

What does it mean to be human? Ultimately, how one answers this question determines not just the course of an individual's life, but also of our collective human civilization, as well as everything non-human—celestial as well as terrestrial—that our collective lives impact. It has long been observed that our collective ecological crisis is the result of many humans considering themselves apart from, rather than part of, the natural world. In non-Indigenous and Westernized cultures, the schism between nature and culture is deeply entrenched in generational cultural patterns that unconsciously embrace anthropocentrism, individualism, and faith in scientific progress—what Chet Bowers refers to as the root metaphors of modernism.[7] Each of these problematic metaphors, along with others, such as racism, patriarchy, and economism, reinforces a stance that sees the natural world (i.e., the biota and the earth systems that support it) as a mere backdrop for the human drama. Nature, or parts of it anyway, becomes something to be used or exploited as a natural resource in the project of human civilization. Under this dominant paradigm, the significance and fate of non-human nature is determined and shaped by human beings.

Death, however, is the great leveller—re-membering us to our shared pre-cultural origin and a shared destiny in a nature that we cannot escape. Although diverse cultural groups, as well as families and individuals themselves, tell diverse stories about the meaning of death, all of us share the same biological fate. We must die. From a biological and evolutionary perspective, death is the precondition for life. We are the leavings of many deaths in a larger process of constant transformation. If one accepts the story of evolution, all of us evolved from the same physical and genetic material. We—humans and non-humans alike, sentient beings and inorganic presences alike—emerged from the same physical materials and energetic processes. Or, as Walt Whitman put it in the opening lines of "Song of Myself": "every atom belonging to you as good belongs to me."[8] Related descendants of stardust and all of our organic and inorganic ancestors, we all equally belong to this planet; we are all linked in common to the mysteries behind its creation—as well as to its own eventual demise. Life is impermanent and full of surprises. In time, the sun itself will collapse and cool to the background temperature of the universe. Long before that happens, Mars will be more hospitable than earth to life as we know it.

What might it mean to allow the biophysical processes of life and death to inform our understanding of our relationship to each other, to non-human others, and to the cosmos itself? The promise of death beckons us with the shared reality of our common fate, as well as our common origin. To ignore and deny death as an intimate part of what it means to be alive is to turn away from the chief aspect of our common humanity and our link to all the planet's beings. The denial of death, and the wish to escape it by prolonging human life indefinitely through science, is directly related to the continued destruction of the planet's non-human life, as well as the systems that support all life on earth. To turn away from death is to hoard life for oneself at others' expense, and to deny the common facts attached to every life, including one's own. Welcoming

death into our beliefs about what it means to live is to embrace the precious-
ness of life and to experience compassion for the losses that everyone, eventually, shares in common. The development of compassion for others and a sense
of our common condition as human beings—these are necessary components
of ecological virtue.

Bringing death out from the shadows into the common light of day, where
it can be marvelled at and accepted, is a radical act of unity and humility. In
openly acknowledging and accepting our impermanence, we signal our own
limitations and vulnerability. Without such acceptance, the denial and fear of
death become psychological projections—distortions of feelings we cannot
face ourselves. Such projections can fuel the egocentric behaviours of adolescent individualism and the ideologies of "social Darwinism" and capitalism in a
socially and ecologically devastating competition of all against all.[9]

As many ecological thinkers have observed, we need a new story. The root
metaphors of modernism—individualism, anthropocentrism, faith in progress, etc.—serve neither humanity nor our shared habitat. To serve life, the
new story must be rooted in metaphors of unity expansive enough to allow
for difference. Embracing the mysteries and certainties surrounding death can
provide us with common ground, as well as foster a sense of compassion for
oneself and an understanding of diverse others.

4.

# Death Literacy Teaches Empathy for the Suffering
## of the Planet, and Compassionate Action

One of the most troubling manifestations of the Anthropocene is our growing
deafness to the language of the natural world.[10] Why are we not filled with sorrow and remorse? Why are we not stirred to empathy and action? How might
we learn to listen to the earth, and what are we to do if we are overwhelmed
by what we hear? Death literacy must include literacy about the suffering of
non-human life on the planet, and about how we might encounter that suffering without looking away or lapsing into pessimism and inaction.

Our capacity for empathy—for "feeling into" the lives and experiences
of others—is an innate gift of mammalian evolution. Anyone who has ever
felt their body stir in sympathetic movement when they watch a dancer, or

witnessed a yawn spread or laughter erupt contagiously through a group of people, knows the power of empathy. It is this same capacity that works in us when we read great literature or watch a movie. We imaginatively feel ourselves into the lives of fictional strangers; we enter experiences and ways of being we had not known before, and we are changed by it. This is the wonder of empathy: by mirroring in our own bodies the emotional world of others, we can get a sense of what they are feeling and thinking, and we can imagine the world from their point of view.

Anyone who has bonded with a pet knows how easily our capacity for empathy can extend to the animals we love. Art exerts a similar magic, inviting us into a kind of conversation that we feel our way into, letting it work upon our understanding of the world and ourselves. The same capacity for empathy can extend to the living wild places and landscapes we come to love. When we stand in their presence, not forcing understanding but rather feeling our way into a bodily-felt empathy, as we would with music and art, waiting and letting nature speak directly to our animal selves, the conversation begins, and we hear her.

The obvious question is this: If empathy comes so naturally to us, why do we seem not to care that our precious planet is being raped and pillaged, that entire species are dying out of existence, that oceans and forests are depleted and sick? Put differently, what happened to our empathy? Why are we turning away?

The truth is, as wired as we are for empathy, and as easily as it seems to flow in us, it also has predictable failures. If we are serious about becoming literate about death, we have to face and understand these challenges and work at overcoming them.

The first challenge is our disinclination to empathize when the "object" of empathy is too different from us—the stranger, for example, or the dispossessed and marginalized who we think of as "other" or "less than us" and therefore unworthy of our care and concern. It is the habitual objectification of "the other" that allows us to distance ourselves and justifies our lack of concern. Homeless people become invisible in exactly this way: when I unconsciously affirm to myself that they are not like me, I look through and around them, and no longer even see them. It is in those spine-tingling moments when we no longer look around but into the person—when we see for the first time *the face*[11]—that our capacity for empathy and compassion is re-enabled. Martin Buber might say this moment of recognition is the moment I see the other as "thou" rather than "it."[12]

Similarly, empathy for the planet and its diverse forms of life has been one of the casualties of our turn toward objectifying and exploiting the earth. Without sacred stories that teach love and respect, and rituals that allow us to enact that love and respect, we no longer relate to her as *thou*, but as *it*. Interestingly, we remember her and pay her homage at some of our most precious rites of passage—weddings, funerals, baptisms, and graduations—with ceremonies that include flowers, food and wine, fire and water. But these vestigial memories of her importance in our lives fade. We forget. And very little about our everyday lives urges us to remember.

Most urbanized North Americans are overly distracted by our social and private concerns. Many people simply do not have the attention span for, or sustained exposure to, the natural world to register empathy or a sense of connection to her. Much of our understanding of the non-human comes not from direct experience and reflection but from technologically mediated commercial sources. Instead of deep, reciprocal relationships developed over time, we have brief, incidental encounters with nature that we often think of as annoyances—contact with weeds, pests, and insects, for example. Our perspectives toward the non-human realm are also sanitized and distorted in places like zoos and cultivated gardens, or in other domesticated places like classrooms, where we learn facts about nature but not how to relate to her as a "thou" rather than an "it." We simply don't see the planet as a living presence to be protected and cared for.

Further, the wilder places that might surprise and soften even the hardest hearts are out of the reach of average city-dwelling families, and even those natural oases that still exist are always in jeopardy of being lost to "development." If we do encounter real wilderness, it is so opposite to the physical environments we spend our lives curating and controlling that we have no frame of reference with which to even begin the conversation. We are like foreigners in a strange country, so unfamiliar are we with the language and customs of the wild. The effect can be so disorienting and upsetting that we end the conversation before it even has a chance to start. That nature seems to have nothing to say to us is not because she has fallen mute; rather, it is a reflection of our unfamiliarity with her languages and our withered capacity to hear her. Many of us have forgotten how to be perfectly still and listen.

Empathy for the suffering of non-human others, and for the planet itself, is desperately needed, but empathy will not arise in us until we can unravel the popular discourses and habits that construct the living planet as an *it*. We need stories, songs, dances, and poems that help us celebrate her as *thou*, and we need ceremonies and rituals to encode those celebrations in our bodily and communal memories. If we don't have them, we need to create them.

In this age of widespread ecological destruction, our lack of empathy is problematic in a second way. Perhaps the biggest challenge comes as we awaken to the damage we have caused: recognizing our culpability and overwhelmed by remorse, we once again turn away. From the experience of emergency responders and health-care professionals working in high-trauma environments, we know that repeated assault on a person's capacity for empathy can leave them traumatized, depressed, and emotionally numb.[13] We should take heed. The cost of really seeing the ecocide we are accountable for will be overwhelming sorrow and remorse, which can deplete our energy and press us into despair. Only by mobilizing the force of our grief into social action will we find a way forward from that place of despair.

Death literacy must awaken and encourage our innate capacity for empathy with all living things, so we are moved to sorrow by their suffering and death. Children barely need instruction; only the opportunity to explore, even if that exploration is limited to city parks, laneways, and sidewalk cracks where weeds grow, or to staying up late to watch the night sky, or reading about the wonders of nature, and watching documentaries about elephants, or galaxies, or trees that talk to each other.[14] Their capacity for empathy will thrive if given half a chance. And it is not too late for the rest of us: if we give ourselves permission to slow down and find a quiet place where the animal body of our emotional selves can reconnect to the earth, and our capacity for awe and wonder can awaken, we will learn to hear the earth again.

We have already spoken of the importance of emotional literacy as a necessary component of death literacy, but if we are wise, we will protect our precious capacity to care and ration our exposure to everything that through repeated bombardment desensitizes us—horror movies about apocalyptic end times, for example, or sensationalized news of doom from around the globe. Pessimism is the enemy of empathy and compassionate action.

Even if our only contact with the earth is the night sky and the sun and rain on our faces, let us accept the invitation to conversation and allow ourselves to feel our way back into relationship. Our precious planet cries out for compassion. We must practise listening, until we become a container that strengthens with time to hold the suffering we hear and that, sooner or later, we will eventually encounter in our own lives.

# 5

# Death Literacy Teaches that Illness
## Is Part of the Human Condition

Illness is often our first taste of our own vulnerability—a preview of the losses and declining abilities that aging brings. In our death-phobic, youth-obsessed culture, this is not a preview we tend to embrace, and people who are ill pay the price. Their stories, which are often about failure, suffering, loss of meaning, and renegotiated identity, are hard to hear. They disrupt the prevailing cultural imperatives to "be well" and to "age well." They are reminders of our common human vulnerability and fate, and so we look away.

We have a tendency to listen according to certain preferred narrative arcs. With illness, and with suffering more generally, we prefer stories that have a happy ending, or at least offer the possibility of recovery.[15] We don't like the messiness of illness and disability. "Get well" cards typify this preferred narrative. We also like stories of heroism in suffering: stories of "battling" cancer, for example, or stories where the ill person finds new insights and a purpose for living. These narrative preferences privilege the comfort of illness-phobic listeners over the transformative possibility of hearing the truth. Subtly but powerfully, through broken eye contact, inattention, and other signs of squeamishness, the listener lets the ill person know when their story is veering away from these preferred narrative arcs, and so transgressing the bounds of acceptable conversation. The result is that the ill person prunes and sanitizes their story, leaving out the unruly emotional disclosures and the truly horrifying aspects of their experience. But the tamed story is no longer a story that emboldens teller and listener with the truth, and neither teller nor listener benefits in the end. Both are diminished. The ill person is exiled to silence, and the listener has squandered the possibility of learning something of immeasurable value: that illness is not the enemy but rather the risk we take in being alive—a risk that none of us will escape.

A robust death literacy must teach the inevitability of human vulnerability and illness. We are meant to enter these stories, as we would enter any good story, ready to encounter what is actually there and open to the possibility of being transformed. Many people who suffer ongoing pain, disability, or illness still experience a sense of wholeness and physical integration when they are connected and held in meaningful relationships to others who listen in this way. The gift to those who listen is the fuller realization that the worth of a life cannot be measured in terms of productivity or any of the other trappings of "success," and that the lives of the ill, the disabled, the old, and the dying have value *as they are*—not as they *might have been* pre-illness or pre-disability, and not as *they might be* at some future moment when the illness has been treated or the loss of function restored—but precisely *as they are*. Since illness and declining ability await us all, this lesson cannot begin too early. The ability to confront illness in ourselves and others—with eyes and hearts wide open—is a prerequisite for noticing and feeling the loss of anthropogenic ecological decline.

6.

# Death Literacy Teaches Us About Absence, Awakening Us to the Presence of Ancestors and Elders

In his classic poem "Thanatopsis" (meaning meditation on death), early American poet William Cullen Bryant imagines the generations of dead entombed in the "mighty sepulchre" of the earth. He reminds us of the fact that "All that tread / The globe are but a handful to the tribes / That slumber in its bosom."[16] This simple fact—that those alive today are dwarfed in number by the multitude of dead—provides some comfort to the poet, who realizes that death simply means sharing the fate of generations of ancestors, joining the ever-expanding family of life in a final release from worldly care.

Despite the popularity of family genealogies and new opportunities for genetic inquiry, most of us live our lives without much thought about the numberless tribes of relatives who endowed us with living. The popular sentiment is that the past is something from which to escape, that we are new and improved. Indigenous traditions that honour the ancestors are an instructive reminder to modern people that all of us have Indigenous ancestors who lived closer to the earth—closer to death and the cosmic-terrestrial ground of our being. Expanding our moral imaginations to encompass our ancestors' earthy incarnation can be a powerful corrective to cultures and psyches addicted to a technological future and out of touch with heritage. As any historian knows, the past is not over. What we think of those who have gone before, whether we think of them at all, in large part determines our fate. To neglect the dead is to neglect the biological and cultural origins of our humanity and to misconstrue life's riddles and mysteries as ours alone.

Chauvet Cave in southern France contains the oldest and best-preserved cave paintings in the world, as well as what may be the oldest human footprints that can be accurately dated. The ancient artwork depicts at least thirteen different species, including horses, bison, mammoths, reindeer, lions, hyenas, panthers, bears, and rhinoceroses. The paintings are between 32,000 and 37,000 years old. Werner Herzog's 2010 film *Cave of Forgotten Dreams* documents a rare excursion into Chauvet Cave, a UNESCO World Heritage Site that has been carefully preserved and sealed off from public view since its discovery in 1994.

The fully realized artistry of Chauvet is a haunting message from the deep past. One of its more striking panels depicts a group of lions staring fixedly together at some unknown subject—possibly human, possibly other-than-human. These early Paleolithic people had minds, bodies, senses, and lives as vital as our own. They had family, community, rituals, art, and, according to Herzog's interpretation of the cave paintings, a reverence for the awesome mysteries behind creation and the cycles of life and death that we share with all beings. These early people left only traces to remind us that we inhabit the same terrestrial realm, under the same sky, the same sun, the same rain. In modern life, ancestry, the unfathomable depth of the past, represents an awesome and paradoxical absence: our very presence is completely dependent on the generations that came before, yet in our own complicated, crowded, and self-absorbed lives, we hardly think of them at all. Such neglect allows us to continue living as if we were our own inventions, as if the question of what it means to be human is only a recent concern: the psychological dilemma of the individual self. The handprints, footprints, and cave paintings at Chauvet remind us otherwise.

Like our ancestors and other extinct species, death represents an absence that is no less present for its place in the shadows. The mind searches in vain for something solid, for presence, for something it can count on. Absence defies this quest for the concrete, for the provable and the known. It forces us into a choice—face mystery or shrug it off, turn toward or away from death.

To honour our ancestors is to include the dead, and death itself, in our conception of what it means to live. To do so is to expand the moral imagination to include within our world view the unprovable mysteries that resist the achievements of scientific discovery. In modern society, nothing is more abhorrent than something that cannot be explained. Establishing a relationship with the past, with the unknowns surrounding the lives who came before us, requires the virtues of curiosity, humility, and an appreciation for knowledge and wisdom we may not even realize we lack. Our elders, those still alive and those who left parts of themselves in the form of writing and other artifacts of human expression, may be our best link to the past, to a resonant absence waiting to be regenerated.

We use the word "elder" with intention. Most contemporary people are not members of communities where elders and ancestors occupy places of the highest esteem. The structure of modern life divides and segregates old from young, past from present, life from death. We lack access to elders, we don't talk to our ancestors, and we fail to offer them attention and respect. We are only diminished by their absence from our lives, and it is an absence that many seek to remedy. Filled with absence, many people now yearn to connect with their forgotten ancestry, to learn more of their family and immigration stories across time and place, to retrace and reconnect with their origins. Many people from all backgrounds also hunger for those nearby who are older and wiser than we are, and who have something vital to share. Elders and wisdom keepers are links to a deeper past than we usually allow ourselves to experience in our haste, or even in our mindful desire to become more fully alive in the present.

What elders can remind us of, if we care to pay attention, is that past wisdom is available now and needs to be remembered into the uncertain future. Barry Lopez, American nature writer, storyteller, and student of Indigenous cultures—himself an elder—writes: "We repeatedly lose touch with what we intend our lives to stand for. To protect us, the elders must constantly reacquaint us with our ideals." Forgetfulness of past wisdom, Lopez says, is the "Achilles heel of human consciousness—the lapse and disintegration of memory."[17] To be literate about death is to recover the virtue of wanting to know what our lives stand for in the deeper sweep of time and to live into that image; it is to remember that in facing life's challenges, including the challenge of dying, we are not alone. We are literally surrounded by accumulating absences that might help guide us over our next threshold.

CODA

# "Remember the Monarchs"

The history of environmental thought and action is peopled with many long-buried mentors, elders, and ancestors. As Ralph Waldo Emerson put it in the early nineteenth century, "there is properly no history, only biography."[18] If we care to look, we might find in the story of past lives models of how to live and how to die.

Rachel Carson is most often remembered as the author of *Silent Spring*, the 1962 classic that arguably launched the modern environmental movement. The book was the first major exposé of chemical pesticides and the corporations that promoted and peddled what she called "the chemical death rain" of industrial agriculture.[19] Much of *Silent Spring* focused on linking toxic chemicals to cancer. It is a sad irony of history that Rachel Carson herself would suffer from advanced stages of cancer and die—just as her bestselling book began to catalyze shifts in both public perception and public policy around the role of toxins in our environment.

But *Silent Spring* is only one part of Carson's legacy. Years before its publication, Carson had already become famous for writing three other bestselling books—all of them about the sea. Each celebrates the miracle of life; they are biographies of the oceans that place human beings within the wondrous story of evolutionary biology and the much larger sweep of geological time. As a biologist, Carson thought often about death. She saw it as a natural part of the life cycle of the organism, and as part of the evolutionary process. As a writer, Carson described the world of life and death in a way that both revealed mysteries and embraced the mysterious. In one of her first major essays, she wrote (describing evolution): "Individual elements are lost to view, only to reappear again and again in different incarnations in a kind of material immortality." Each life, she says, is not "a drama complete in itself," but only "a brief interlude in a panorama of endless change."[20]

It was Carson's sense of wonder for the natural world that motivated her work as a writer, scientist, and citizen determined to defend life against the "elixirs of death."[21] At the end of her life, when she was quite ill and living in constant pain, Carson wrote to her closest friend about her own impending death. In her letter, she recounts a recent visit to their favourite place together on the Maine coast:

For me it was one of the loveliest of the summer's hours, and all the details will remain in my memory: that blue September sky, the sounds of the wind in the spruces and surf on the rocks, the gulls busy with their foraging, alighting with deliberate grace, the distant views of Griffiths Head and Todd Point, today so clearly etched, though once half seen in swirling fog. But most of all I shall remember the monarchs, that unhurried westward drift of one small winged form after another, each drawn by some invisible force. We talked a little about their migration, their life history. Did they return? We thought not; for most, at least, this was the closing journey of their lives.

But it occurred to me this afternoon, remembering, that it had been a happy spectacle, that we had felt no sadness when we spoke of the fact that there would be no return. And rightly—for when any living thing has come to the end of its life cycle we accept that end as natural. For the Monarch, that cycle is measured in a known span of months. For ourselves, the measure is something else, the span of which we cannot know. But the thought is the same: when that intangible cycle has run its course it is a natural and not unhappy thing that a life comes to an end.

That is what those brightly fluttering bits of life taught me this morning. I found a deep happiness in it—so, I hope, may you.[22]

"Remember the monarchs": These immortal words are etched into a stone marker near the place where Carson's ashes were spread to the sea wind. They describe a perspective toward death that is at once rational and scientific, as well as metaphysical and mysterious. Carson's orientation toward death is an orientation toward life. Her courage—her "deep happiness" in the face of her own demise—is but one lesson from our collective ancestral past that might inform a literacy of death for the living.

## Notes

1   W. Whitman, *Leaves of Grass*, ed. Mark Van Doren (New York: Penguin, 1982), 94.
2   H. Aspiz, *So Long! Walt Whitman's Poetry of Death* (Tuscaloosa: University of Alabama Press, 2004).
3   Whitman, "Song of Myself," in *Leaves of Grass*.
4   P. Shepard, *Nature and Madness* (Athens: University of Georgia Press, 1998).

5   For example, P. Kalanithi, *When Breath Becomes Air* (New York: Random House, 2016); O. Sacks, *Gratitude* (Toronto: Knopf Canada, 2015).

6   On Compassionate Communities, see A. Kellehear, *Compassionate Cities: Public Health and End-of-Life Care* (New York: Routledge, 2005). More information on Dying Matters and Groundswell can be found, respectively, at http://www.dyingmatters.org/ and http://www.thegroundswellproject.com/.

7   C. Bowers, *The Culture of Denial: Why the Environmental Movement Needs a Strategy for Reforming Universities and Public Schools* (New York: SUNY Press, 1997).

8   Whitman, *Leaves of Grass*, 32.

9   Shepard, *Nature and Madness*.

10  T. Berry, *The Dream of the Earth* (San Francisco: Sierra Club Books, 1988).

11  E. Levinas, *Entre Nous: On Thinking-of-the-Other* (New York: Columbia University Press, 1998).

12  M. Buber, *I and Thou*, trans. Walter Kaufmann (Edinburgh: T.&T. Clark, 1970).

13  See, for example, C.R. Figley, ed., *Compassion Fatigue: Secondary Traumatic Stress Disorders from Treating the Traumatized* (New York: Brunner/Mazel, 1995), 1005.

14  P. Wohlleben, *The Hidden Life of Trees: What They Feel, How They Communicate Discoveries from a Secret World* (Vancouver: Greystone Books, 2016).

15  A. Frank, *At the Will of the Body: Reflections on Illness* (Boston: Houghton Mifflin, 1991), and Frank, *The Wounded Storyteller* (Chicago: University of Chicago Press, 2013).

16  W.C. Bryant, "Thanatopsis," Poetry Foundation, https://www.poetryfoundation.org/poems-and-poets/poems/detail/50465 (accessed July 1, 2017).

17  B. Lopez, "Introduction to 'I, Snow Leopard,'" *Orion Magazine*, July-August and September-October, 2016, 85.

18  R.W. Emerson, *Selected Essays*, ed. Larzer Ziff (New York: Penguin, 1982), 153.

19  R. Carson, *Silent Spring* (Boston: Houghton Mifflin, 1962), 12.

20  R. Carson, "Undersea," *Atlantic Monthly* 78 (1937): 67.

21  Carson, *Silent Spring*, 15.

22  M. Freeman, ed., *Always, Rachel: The letters of Rachel Carson and Dorothy Freeman, 1952–1964* (Boston: Beacon Press, 1995), 467–8.

*Part III*

# INSIGHTS *from the* CONTEMPLATIVE WISDOM TRADITIONS

# FIVE

# What Are "Daoist" Virtues?

## Seeking an Ethical Perspective on Human Conduct and Ecology

PAUL CROWE

## Preliminary Concerns

THIS COLLECTION OF ESSAYS CONSTITUTES A CALL TO REFORM our ethics by considering alternative possibilities taken from a variety of cultural sources. This chapter proffers resources embedded within texts and traditions associated with and subsumed under the somewhat bemusing term "Daoist."[1] Prior to opening that discussion there are some preliminary concerns to address. We must acknowledge that anyone advocating a deep and pervasive change in perspective on the relationship of human beings to the ecological fabric that sustains them faces a daunting challenge.

Broadly speaking, the project undertaken in this book is one of advocating for a fundamental reorientation: firstly, in how we conceive of personal ethics, and secondly, in shifting away from the central and abiding concern of ethics as they have evolved in a European context, largely focusing on inter-human conduct and the need to secure co-operation within human communities in the interests of providing the requisite mutual safety and associated freedom to pursue our lives in accordance with private aims. These latter central concerns are starkly evident in the work of Thomas Hobbes, and in that of John Locke, who inspired the language of the American Declaration of Independence and the best-known set of unalienable rights listed in that document: "Life, Liberty and the pursuit of Happiness." Of course, after decades of indoctrination aimed at normalizing our status as consumers rather than fully human beings, coupled with the moral imperative to consume, the final word in that famous phrase could now be replaced with "stuff." What challenges present themselves to those of us who would wish such a tectonic shift in the moral landscape for the sake of our very survival as a species? How might we define our moral responsibilities, not only to other human beings but to the very biosphere that embraces, nourishes, and constitutes us? How are we to persuade people that such a change is in their best interests?

A significant challenge is the widespread perception that wealth as stock, and income as facilitator of consumption flow, on both national and personal scales, are positively correlated with a sense of personal well-being, life satisfaction, or happiness. No politician dare suggest an urgent need to attenuate our current standard of living. No minister of finance in the G20 advocates for mere maintenance of, let alone the slightest contraction of, their national economy. The theology to which such figures must adhere is one of perpetual growth by harnessing greed to fuel consumer spending. In support of arguments for alternative values, perceptions, and assumptions about the nature of the world, and the place of *Homo-sapiens sapiens* in that world, it is tempting to point out that in fact the association of happiness with capital accumulation and increased income is a false one. Unfortunately, despite the popular cliché that you cannot buy happiness, the opposite is borne out by what appears to be credible research.[2] Further, the claim that there are diminishing returns on life satisfaction as wealth increases—the so-called satiation hypothesis—is also far from self-evident:

> Bringing together a number of international survey data-sets that cover about 90 per cent of the world's population, including many developing countries, Stevenson and Wolfers (2008, 2013) ... found that "the relationship

between well-being and income is roughly linear-log and does not diminish as incomes rise," and concluded ironically that "if there is a satiation point, we are yet to reach it."[3]

Senik goes on to conclude, "there is no consensus on the existence of a threshold of wealth beyond which the marginal utility of income would fall to zero."[4] Thus, the first difficulty is convincing people that they should give up on a mode of economic activity that provides immediate assurance of, or at least the hope for, future happiness. This may actually require tolerating a measure of sacrifice.

Personal sacrifice is linked to a second challenge: the bizarrely contradictory attitudes that people hold concerning "the environment" in relation to their own levels of life satisfaction. A 2016 Pew report states that 75 percent of Americans claim to be "particularly concerned about helping the environment," and yet the same report notes that only 20 percent of Americans say they live in ways that help protect the environment.[5] In his review of research findings concerning attitudes and behaviour associated with climate change, sociologist Martin Patchen observes that "there is consistent evidence that people's willingness to take specific environmentally-helpful actions or support specific pro-environmental policies declines as the amount of sacrifice connected to the action or policy increases."[6] Indeed, the tolerance of sacrifice is discouragingly low.[7] Given the dire consequences of mass starvation, water shortages, and insecurity associated with unprecedented human dislocation, the minor inconveniences people are unwilling to endure, such as carpooling, seem staggeringly inconsequential.

These observations are doubly discouraging because they refer to surveys of people who admit that there are serious ecological problems caused by human activity yet feel woefully unmotivated to take action.[8] To say that people fail to act due to successful misinformation campaigns sponsored by corporate interests among large oil, gas, and coal mining companies falls short of the complete truth. Even if people admit the problems, they remain unmotivated to take serious and immediate action.

These points are made partly to acknowledge that this chapter is written from a less than optimistic standpoint. Since the conclusion of the Second Word War we have moved from Keynesian economics, which despite its dismal focus on greed and the individual consumer maintained a role for human beings in managing economies through government intervention in capital markets. Since the rise of Friedrich Hayek's insights, and their subsequent amplification by Milton Friedman, that there is little or no need for the state to carefully constrain and manage market forces, we have witnessed a dramatic turn. This began during the Reagan and Thatcher years and continued across party lines in America through the Bush years, and those of Clinton and Obama, and may be about to take an even more vicious turn with the ascendency of Trump. It is becoming increasingly difficult to think outside of this pervasive universe of discourse. Is there yet hope?

There has been progress. Slavery, for example, is now considered "repugnant." Further, denials of dignity and rights based on race, class, and gender are tolerated far less than they were only a few decades ago.[9] The battles fought to achieve these shifts in perception were often fought at the "grassroots" level, with heroic individuals playing a powerful role in instituting change. Attitudes around smoking shifted dramatically from the 1990s, when manufacturers could still advertise on television and associate the inhaling of more than seven thousand toxic chemicals[10] with the great outdoors and a "healthy lifestyle." In Canada, attempts to reduce the number of impaired driving incidents have also yielded success, with Statistics Canada recording a 65 percent reduction in 2015. These were the lowest rates since records started in 1986.[11] How does this happen?

In his new book *The Myth Gap*, Alex Evans suggests that compelling stories appear to be more powerful than carefully detailed scientific evidence and associated statistical data. The election of Trump in the United States and the success of the Brexit campaign in the United Kingdom are examples of how misinformation and an absence of facts can win out if framed within a sufficiently compelling narrative.[12] Conversely, Evans points to success in the Paris climate talks, with the declaration of zero emissions by 2050, as resulting from a complete reversal in tactics by leaders in the environmental movement, who realized that data alone will not triumph in the quest to win broad public and political support.[13] Thus, there may be hope for change if we can develop faith in a compelling new myth to replace what Ronald Wright summarizes as "the idea that the world must be run by the stock market," which he concludes "is as mad as any other fundamentalist delusion, Islamic, Christian, or Marxist."[14] Perhaps the requisite elements of this new story can be discovered in a variety of cultural outlooks for the purpose of creating a compelling new myth to replace the old one of humans set against an inert natural backdrop ripe for exploitation. Before we consider the kind of elements Daoists might add to a compelling and urgently needed story, we need to understand the nature of the cultural domain to which we are directing our attention.

## What Do We Mean by "Daoist"?

Before we can consider contributions to this new story from Daoist communities, we need to clarify what, exactly, we are referring to when we invoke the terms "Daoism" and "Daoist."[15] "Daoism" is an *ex post facto* construction rather than a historical given with a clear point of origin. The first socially identifiable community labelled by contemporary specialists as "Daoist" did not appear until the close of the Han dynasty (206 BCE–221 CE). Leaders of this theocracy in western China, occupying part of the region now known as Sichuan, did not call themselves Daoists. They adopted the name Celestial Masters (Tianshi 天師). It was their aim to usher in a new age of Great Peace (Taiping 太平) during a period of political instability, violence, dislocation, and widespread disease. The Celestial Masters' community disintegrated in the years of warfare and realignments of military and political power commencing with the collapse of Han central authority. Subsequent centuries saw the rise of a variety of approaches to cultivation by several major movements that blended a variety of teachings in unique ways and claimed their own identities. None identified themselves simply as Daoists. Instead, they identified themselves with the teachers, texts, and outlooks on cultivation that informed their lives.

The modern term "Daoism" is derived from two classical Chinese terms: *Daojia* 道家 and *Daojiao* 道教. The former began as a bibliographic category established by Han dynasty court astrologer Sima Tan 司馬談 (d. 110 BCE).[16] He attempted to organize a variety of approaches to effective government into six abstract categories, three of which he coined, including *Daojia. Jia* can mean "family" and by extension "school," though here as a purely abstract category.[17] The *jiao* in *Daojiao* as a noun means "a teaching" and, as a verb, "to teach." Thus, *Daojiao* means "teachings of the Dao." The significance of Dao and Daoism shifts with context and is subject to interpretation. Attempts to essentialize it in the form of *a* religion or *a* philosophy are doomed to be both ahistorical and imprecise. Thus, any account of a body of virtues or general orientation toward the natural world within which we exist must account for the rich diversity of individuals and communities recounted above.

## What Are Daoist Virtues?

Given the diversity of approaches to personal transformation associated with the various groups now denoted by the term "Daoism," what elements might we productively harness to address our present ecological plight? While not reducible to some essential core, there is nonetheless a range of insights that constitutes a sphere of family resemblance. A further complication is that this set of qualities and values have had a fluid relationship with those sharing the same geographic and historical coordinates but not identified as Daoist. Ru literati and Buddhist ideas concerning self-cultivation and virtue cannot easily be partitioned off and will be factored in to this discussion.

The term "virtue" as understood by the figure of Socrates in Plato's dialogues, and by Aristotle, points us generally in the direction of characteristics associated with personal excellence (*aretê*) and human flourishing (*eudaemonia*). Some of the earliest and most influential uses of *de* 德 by the above groups, usually translated as "virtue," lend themselves quite favourably to an association of this trait or capacity both with a kind of excellence or effectiveness and with human flourishing, though not confined by a teleological characterization. By the Warring States period (475–221), *de* is also understood within the *Lunyu* (Analects of Confucius) and the *Daode jing*, as an internal capacity that develops within an accomplished individual. This potency affects change without an individual needing to take overt coercive action. Thus, in the *Lunyu* an ideal leader is comparable to the North Star, which simply maintains its position in the sky as the other stars gather around it.[18] This approach is later echoed in the *Daode jing*, where superior virtue is described as the type of effectiveness emerging out of detachment from purposive, narrowly defined action and denoted by the term *wuwei* 無為, which, though a central notion among groups now designated Daoist, was first used in the *Lunyu*.[19] Thus, as with the North Star, it is a matter of simply exuding a kind of influence over circumstances associated with the person of a leader or sage.

How is virtue cultivated and how is it conceptualized? Early conceptions of this "virtue" are provided in the *Neiye* 內業 (Inner Achievement), dated ca. 350–300 BCE and constituting chapter 49 of the *Guanzi* 管子. The *Neiye* was continuous with fundamental presuppositions about virtue that would later appear in the *Daode jing* and which also appear in the classic fourth-century Ru text known as the *Mengzi* 孟子 (Latinized as Mencius). The *Neiye* describes a clear link between virtue (*de* 德), wisdom (*zhi* 智), and the *qi* 氣 or *jing* 精[20] circulating through the body and through the cosmos as a life force:

> The vital essence of all things:
> It is this that brings them to life.
> It generates the five grains below
> And becomes the constellated stars above.
> When flowing amid the heavens and the earth
> We call it ghostly and numinous.
> When stored within the chests of human beings,
> We call them sages.[21]

> 凡物之精, 此則為生下生五穀, 上為列星. 流於天地之間, 謂之鬼神, 藏於胸中, 謂之聖人.[22]

The "vital essence" (*jing* 精) is defined in the eighth verse of the *Neiye* as being an essential form of life force (*qi* 氣).[23] As such it can be distinguished from the less pure form ordinarily circulating throughout the human body. If this vital force can be drawn into the body one becomes a sage—that is, a person of robust health, high accomplishment, and true wisdom. It is this embodied vital force that gives rise to and sustains virtue (*de* 德). This force also binds together and permeates all three divisions of the world: the sky above our heads, the earth beneath our feet, and the bodies they sustain. Thus, in a very concrete (though rarefied) sense, our efficacy as human beings and our reserve of wisdom depend upon our link to the sky and earth.

Embodied and cosmic dimensions of virtue and wisdom are evident in another text of the same period: the *Mencius* (Mengzi), mentioned above and associated with the Ru ("Confucian") teacher Mengzi (372–289 BCE), contains a passage in which this type of assumption is explained in some detail. In a conversation with Gongsun Zhou 公孫丑, Mengzi is asked what his strengths are and he says that he is good at nourishing his "flood-like *qi*" (*haoran zhi qi* 浩然之氣). Gongsun Zhou asks what he means. Mengzi explains that this *qi* is expansive and that if fed by integrity (*zhi* 直)[24] and not injured in any way it will fill up the space between the sky (head) and earth (feet). This *qi*, claims Mengzi, links appropriate human conduct (*yi* 義) to Dao. Through this medium of *qi*, human action can be attuned so that it resonates with and is informed by Dao. As with the *Neiye*, there is an assumption of a tripartite universe within which the body of human beings is linked through the fundamental life force of *qi* to both sky and earth. Further, once again it is assumed that human conduct is the catalyst for this connection. Mengzi makes the point that the "flood-like *qi*" is not nourished as a result of incidentally appropriate acts. Instead one's moral actions must be continually attuned in such a way that they are appropriate within any given context:

> It is a Ch'i [*qi*] which unites rightness and the Way [Dao]. Deprive it of these and it will starve. It is born of accumulated rightness and cannot be appropriated by anyone through a sporadic show of rightness. Whenever one acts in a way that falls below the standard set in one's heart, it will starve.[25]

> 其為氣也, 配義與道; 無是, 餒也. 是集義所生者, 非義襲而取之也. 行有不慊於心, 則餒矣.

The final phrase indicates that appropriateness (*yi* 義) is not measured deontically through adherence to an external moral standard or set of rules, but principally through a "standard set in one's heart"—in one's body. If it falls short of this standard, then the *qi* will starve. Since it is the same *qi* coursing through our bodies and through the larger cosmic body within which we are subsumed, the stability of our link with the wider forces of the world around us is intimately tied to our moral conduct. Another assumption in these texts is that human health is contingent upon the strength and viability of the *qi* in our bodies and the viability of that organic connection with the wider world. Thus, morality and health are collapsed into a unified cosmo-physiological vision.[26]

Links between the human body and one's moral comportment are very literal in both the *Mengzi* and the *Neiye*. The *Mengzi* includes the following observation:

When he is upright within his breast, a man's pupils are clear and bright; when he is not, they are clouded and murky. How can a man conceal his true character if you listen to his words and observe the pupils of his eyes?[27]

胸中正, 則眸子瞭焉; 胸中不正, 則眸子眊焉. 聽其言也, 觀其眸子, 人焉廋哉?

And again,

That which a gentleman follows as his nature, that is to say, benevolence, rightness, the rites and wisdom, is rooted in his heart, and manifests itself in his face, giving it a sleek appearance. It also shows in his back and extends to his limbs, rendering their message intelligible without words.[28]

君子所性, 仁義禮智根於心. 其生色也, 睟然見於面, 盎於背, 施於四體, 四體不言而喻.

This Ru literati expression of embodied moral dispositions illustrates that what might be considered a uniquely Daoist perspective was already a more pervasive assumption.[29] Echoing the *Neiye*, the *Daode jing* (verse 53) observes that those who embrace virtue become like infants with bodies that are soft and supple with strength and great endurance. Further, the inner harmony associated with this state supports constancy and wisdom.[30] Some five centuries later we find these ideas reproduced during a renewal of Ru political thought bolstered by cosmological speculation.

With the chaos at the end of the Warring States and the subsequent collapse of the Qin dynasty (221–06 BCE), the first emperors of the succeeding Han dynasty (206 BCE–220 CE) worked to centralize power. There was considerable interest during this period in systems of thought that provided a cosmic justification for the transition to the Han and consolidation of central authority vested in the emperor. An important figure often credited with this development is Dong Zhongshu 董仲舒 (195–05).[31] He described the larger world as a kind of organism, with the *qi* of the sky above, the *qi* of earth below, and the *qi* of human beings occupying the middle space between them. He also believed that the behaviour of human beings, and particularly the emperor, had a direct effect on the sky and earth because of a shared resonance. He predicated this resonance on the likeness between human beings and the world in which they lived. Thus, in comparing sky and earth to human beings, he states,

> Human beings have three hundred and sixty joints, which matches the sky's number [of days in a year].
>
> Our bodies have bones and flesh that matches the fullness of the earth.
>
> Above we have keen hearing and sight of ears and eyes, which resemble the sun and moon [in the sky].
>
> Our bodies have orifices and veins, which resemble rivers and valleys [on the earth].

> 人有三百六十節偶天之數也
> 形體骨肉偶地之厚也
> 上有耳目聰明日月之象也
> 體有空竅理脈川谷之象也[32]

Dong thought that this resemblance entailed a kind of resonance akin to sympathetic vibrations associated with music. He supported this idea by pointing out the sonic resonance between a *guqin* 古琴 and a standing harp called a *se* 瑟. When the *gong* 宮 note (first in the pentatonic scale) is struck on one instrument, the corresponding string vibrates on the other, and when the *shang* 商 note (second in the pentatonic scale) is struck, the same note sounds on the other instrument.[33] These ideas also gave structure to speculations in the field of medicine that began to take on theoretical coherence during the later Han dynasty. Dong Zhongshu understood that this resonance between human beings and the sky and earth had moral implications. Poor behaviour disrupted the harmony between the human realm and that of the cosmos, and the results of such disruptions could be seen in celestial and terrestrial anomalies.[34] Serious consideration was given to belief in the close mutual relationship between the wider natural processes and the lives of human beings. At the highest stratum of society, court proceedings were dictated down to the finest details, such as the colour and dimensions of court regalia and the length of court documents, and by the observation of ritual rules dictated by correlative connections. Dong placed the emperor at the centre of this system; it was the emperor who bore the burden of ensuring the smooth functioning of the cosmos through regulation of his conduct in accordance with the recommendations of ritual specialists. Nearly four hundred years later, as the Han fell into decline during the closing years of the second century amid rebellion, internal strife, and natural disasters, the central role of the emperor in facilitating harmony was augmented by a vision expressed within the Celestial Masters theocracy in western China.

To the Celestial Masters, the flow of *qi* circulating between sky and earth sustained the world and nourished human beings. Just as hindrance to blood flow in the body brings pain, illness, and eventually death, so interference with the animating flow of *qi* through the cosmic body brings illness. Sinful behaviour was one root of such destructive forces, as it led to a disruption of human society that rippled through to the natural world; natural disasters, illness, and starvation were seen as evidence that previous generations had accumulated bad conduct that was now being felt by contemporary society. Bad conduct, like viruses, had communicable and long-term effects. One aspect of conduct involved human treatment of the earth and sky, or, as we might say, nature or ecology. This notion that we transmit our present ecological sins to future generations is of course a fact we need to consider seriously when we reflect on the consequences of our ecologically disruptive and destructive behaviour.

Human beings continued to be understood as an integral part of something akin to a larger cosmic "organism." However, amid the Celestial Masters' community, the general familial context central to Ru social and moral thought, with its associated educational and nourishing functions, bound by filial obligations and relationships, are superimposed on what is conceived of as a cosmic family: human beings owe a filial debt of respect and care to their parents, the sky-father and earth-mother. After all it is earth-mother that provides nourishment to her children while sky-father provides order and rhythm to life. Failure to honour the filial dimension of our relationship to earth and sky has terrible destructive ecological ramifications.

We gain insight into the Celestial Masters' world view from one of the longest texts in the Daoist canon, the *Taiping jing* 太平經 (Scripture of Great Peace)[35] (CT1101; HY1093).[36] Of the 170 chapters thought to have constituted the original, there are fewer than 60 remaining in the longest extant version. Here, a section of chapter 45, titled "Detailed Explanations on Raising the Earth and Setting Forth Texts" (Qitu chushu jue 起土出書訣), is used to describe the relevant details. The chapter recounts a conversation between a disciple and one of the Celestial Masters. The disciple is concerned to learn about the pain he was sure human beings cause the earth and the repercussions this behaviour entails. The student begins in standard deferential form:

> [I am] a lowly, stupid worthless student who cannot overcome what my heart desires to ask. Violating the Celestial Master's prohibitions is an error exceedingly grave.... A small man [I] cannot endure my feelings and desires. [My] five internal [organs] have developed anxiety and sorrow and I am genuinely ill at ease. May I ask [concerning my] single great doubt, if only the Celestial Master would now treat [me] as [if] classed as a new-born;

> 下愚賤生不勝心所欲問，犯天師忌諱為過甚劇…，小人不忍情願，五內發煩懣悃悃，請問一大疑唯天師既待以赤子之分[37]

The teacher then responds by urging the student to speak, after which the student, referred to by his teacher as True Man, even more humbly begs for instruction on the relationship between humanity and the earth and sky, and, so, the teacher speaks:

> True person, simply sit calmly and listen clearly. What greatly pains sky and earth is evil people failing to go along with [them] and failing to be filial.
>   Now the balance and harmony of sky and earth is entirely [composed of] three qi and they are joined together as a single family;
>   …Furthermore, together they regulate life and together they nourish the ten thousand creatures. Sky directs life and is called father. Earth directs nourishing and is called mother. Human beings direct and regulate patterns [around them] and are called children [or "sons"].

> 真人但安坐明聽，天地所大疾苦，惡人不順與不孝
> 夫天地中和凡三氣，內相與共為一家
> 反共治生共養萬物，天者主生稱父地者主養稱母，人者主治理之稱子[38]

Here we have a vision of the natural world that combines a belief that all things are constituted and sustained by qi with the classical Ru notion of family at the centre of a person's place in the world and their constitution as a person. This allows the Celestial Master to tell his student that human beings, as children of sky and earth, owe them a filial obligation:

As for the child, [its] life receives its destiny from the father and is nourished by the mother. As a child then it ought to reverently serve its father and love its mother.

子者生受命於父見養食於母, 為子乃當敬事其父而愛其母

The teacher then continues by explaining what happens when human beings, as the children, behave in ways that fail to honour the celestial and earthly parents:

Sky nourishes human life and earth nourishes human form; if humans are stupid and ignorant and do not realize how to honour their father and mother they constantly cause heaven and earth [in having] given birth to humanity to have regrets and anxieties they cannot escape.

天者養人命, 地者養人形, 人則大愚蔽且暗, 不知重尊其父母, 常使天地生凡人, 有悔悒悒不解也

The disciple persists, asking the teacher what he means and how human beings cause these regrets, to which the Celestial Master responds,

Presently people take the earth to be their mother and obtain clothing and nourishment from her. [Yet] do not together love and benefit her but instead together deceive and harm her. … People are very insolent and together bore holes into the earth and on a grand scale commence the labour of raising the earth without employing principles of the Dao.

The deep holes descend to the Yellow Springs while the shallow ones go several tens of feet [deep]. The mother is alone, anxious, and enraged while all her children are greatly careless and unfilial.

Constantly embittered and truly unhappy [she] lacks family able to understand her words.

今人以地為母得衣食焉, 不共愛利之反共賊害之... 人乃甚無狀共穿
　　鑿地, 大興起土功不用道理
其深者下著黃泉淺者數丈, 母內獨愁恚諸子大不謹孝
常苦忿忿悒悒而無從得通其言

The teacher goes on to describe how humans act rashly and, unrepentant, deny that they are causing harm to their mother, the earth, explaining that "the earth suffered sickness and pain beyond measure" and was "distressed and anxious that her children could not be regulated." Without recourse, the mother, unable to bare the pain any longer, finally implored the father, the sky, to help her:

> Distressed and anxious that her children could not be regulated above she complained about the people to the father. Her complaints about them accumulated for a long time; therefore the father's anger was ceaseless and myriad beginnings of calamities and strange [occurrences] simultaneously arose.
>
> The mother still did not influence [them] and in constant anger was unwilling to [expend] effort in feeding the people and myriad creatures. Father and mother were both unhappy.
>
> Myriad creatures and people died; they failed to employ the principles of the Dao; herein lies the blame.

> 愁困其子不能制, 上愬人於父愬之積久復久積數, 故父怒不止, 災變
> 怪萬端並起
> 母復不說, 常怒不肯力養人民萬物, 父母俱不喜
> 萬物人民死不用道理咎在此[39]

The Celestial Master is offering an explanation for the terrible woes people were suffering due to political and social turmoil: dislocation, deprivation, disease, and malnutrition. This account goes well beyond previous notions of blame as resting squarely on the shoulders of the emperor. According to the Celestial Masters, humanity itself was seen to be out of step with the Dao and thus incapable of hearing the rhythms of the natural world, just as a rebellious, selfish child is unable or unwilling to hear the desperate pleas of a mother who, through the pain of childbirth and sacrifices made to nourish and support the child, grows weaker and sicker each day. This is a very potent form of language that seamlessly weaves together a Celestial Master view of the natural world filled with resonances and correlations embodied in a cosmic organism alive with three modes of *qi*, with classic Ru literati language centred on the family as the foundation of human cultivation, civilization, and political order.

The teacher goes on to say that not only is there a lack of empathy for the natural world, but indeed humans harbour resentment as though they were not being given enough. Thus, they dig wells with large diameters and dig so deeply that they reach all the way down to the Yellow Springs, where the souls of the dead were thought eventually to reside. Further, people build houses that are overly large and force huge columns deep into the ground. He describes how people dig into the mountains to extract metals and stone and how they block streams and excavate them, which from this perspective, is to block the blood flow of the earth. If people followed the Dao, they would let the water flow freely as it should. The teacher explains that stone is the bones of the earth and soil its flesh; thus, care must be taken to be sensitive to the pain we cause the earth when quarrying, drilling, or using water.

It is important to appreciate that the Celestial Masters were not opposed to human settlement or gaining shelter and nourishment from nature. They believed the earth takes joy in nourishing her offspring. The Celestial Master says:

> Earth is mother to the myriad creatures and takes joy in nourishing them just like a mother with a child in her womb.
>
> If people keep to the Dao and do not recklessly drill into their mother then their mother will suffer no sickness.

地者萬物之母也樂愛養之不知其重也, 比若人有胞中之子
守道不妄穿鑿母母無病

The challenge was to live in moderation, in tune with the Dao or the regular course of nature. People must build houses, but they should not go to excess and build huge structures. It is essential to plant crops and dig wells but take care not to dig too deeply or without regard for the fabric of nature in which such work takes place. By attuning ourselves to nature's rhythms, we avoid becoming terrestrial parasites and instead assume a symbiotic relationship with our parents, sky and earth, one informed by respect, filial piety, and gratitude. In doing so we ensure health far beyond the fragile and limited confines of our bodies and secure an internal peace, balance, and stability that permits us to behave with moral rectitude and regard for our fellow human beings and the natural world that sustains us. It is our connection with the natural world that instills this peace of heart within our bodies and lives and, by extension, the earth.

We find this advocacy for moderation, for conservation, and for treading lightly upon the earth in a much later text attributed to the aforementioned Wang Chongyang (1113–70), the founding figure of Quanzhen dao (the Way of Complete Reality). The *Chongyang lijiao shiwu lun* 重陽立教十五論 (CT1233, HY1223) (Chongyang's fifteen-verse discourse on establishing the teaching), is an introductory manual for novice students containing some of the most basic teachings. The fifth discourse opens with an injunction to embody an attitude of reverence and respect for one's surroundings when constructing a shelter in which to protect and train the body:

A thatched hut or straw shack must cover the body. To sleep outdoors closing one's eyes in the wilds offends the sun and the moon. Consider carved beams and lofty eaves, these too are not within the scope of a highly accomplished master. Audience halls and halls with high ceilings, could these be the handiwork of people of the Dao?

茅庵草舍, 須要遮形, 露宿野眠, 觸犯日月. 苟或雕梁峻宇, 亦非上士之作為; 大殿高堂, 豈是道人之活計.

The verse continues by adopting language strikingly similar to that of the *Taiping jing*:

To hack away at and fell trees is to cut off the life-fluid in the veins of the earth. Turning the Dao into a means for gaining wealth is to take from the veins of the people.

斫伐樹木, 斷地脈之津液; 化道貨財, 取人家之血脈.[40]

The text collapses the distinction between one's conduct toward the earth and other human beings into the single wrong of exploitation—taking with no concern for that from which one takes. There is a clear sense that moral conduct includes both the social relationships and the extended ecological relationships that create, shape, and sustain us. As with the Celestial Masters, what we consider the moral sphere is expanded radically to dissolve a clear distinction between the human and the cosmic.

During Wang Chongyang's lifetime, and that of later generations of his disciples, inner alchemical or Golden Elixir cultivation continued to develop and to incorporate Ru and Buddhist insights into self-cultivation. Alchemy is the process of transforming one base substance into another more refined and precious substance. In the case of "inner" alchemy, the raw ingredients are not metals, chemicals, or plants, but rather the forms of *qi* flowing through one's body. The aim is to transform the individual from a base form mired in the red dust of the world into a person who is in but not of the world, often expressed through the Buddhist metaphor of a lotus flower whose roots are in the mud but whose blossom faces the sky and the sunlight; achieving this state was akin to achieving true immortality. Wang had no time for those who believed immortality meant the indefinite continuity of the body. Rather, the process was about insight and taming the wild emotions of the heart so that one could become like Mount Tai through all the hours of a day.[41] Elsewhere, he explicitly acknowledges the insights of Mencius many centuries earlier concerning the flood-like *qi*.[42] Wang made the same basic assumptions about the material linkage between moral conduct and the body. Further, Wang's assumption that this *qi* renders us part of a larger process of creative transformation (*zaohua* 造化) includes the need to realize the human potential to become a link or conduit between the sky and the earth.

A final figure worthy of mention is Li Daochun 李道純 (fl. ca. 1288), an inner-alchemy master who taught in the Jiangnan region during the Mongol Yuan dynasty (1271–1368) under the reign of Kublai Khan (r. 1260–94). In synthesizing Buddhist and Ru teachings with his own, Li embraced the insights of Confucius and Mencius concerning moral conduct and those of the later *Zhongyong* 中庸 (Maintaining the Centre in Daily Affairs) traditionally ascribed to Confucius's grandson. Though the *Zhongyong* is only a relatively short text within the larger *Liji* 禮記 (Book of Rites), it exercised considerable influence on revivalists of literati culture culminating in the Ru orthodoxy, shaped principally by Zhu Xi 朱熹 (1130–1200). While Mencius had spoken of the flood-like *qi* nourished by appropriate moral conduct, the *Zhongyong* added to this the important idea of inner balance and harmony as a dynamic state responding to the proportionate and appropriate expression of basic emotions. It was through a stable inner life that one could become attuned to the dictates of the wider natural order denoted by the Chinese character *tian* 天, which can be translated as "sky" (especially when used in conjunction with *di* 地 or "earth"). The significance of "sky" needs to be fully understood as entailing the vast process of cyclical change we witness in the heavens through the motions of the stars, sun, and moon, and through the continuous oscillation of the seasons. Thus, while "sky" is an accurate translation on a literal level, it is best to understand *tian* as indicating the natural order of which humans are a part. Li incorporated into his own teaching this idea that we can achieve an internal balance (*zhong* 中) through harmonious functioning of our emotional states. He took the notion of balance or being centred (*zhong* 中) and harmony (*he* 和) directly from the *Zhongyong*. An indication of the great importance he attached to these insights is the fact that he took these terms as the title of one of his major works, the *Zhonghe ji* 中和集(CT249, HY248) (Anthology on the Centre and Harmony). The following quotation appears in the opening pages of the *Zhonghe ji*:

> The *Book of Rites* says: "When joy, anger, sadness, and happiness have not yet come forth, this is called centred; having come forth, if they are all proportioned this is called harmony."

> 禮記云: 喜怒哀樂未發, 謂之中, 發而皆中節, 謂之和.[43]

Li took this state of internal balance and harmony as essential if one is to act in accordance with one's inner nature (*xing* 性), which serves as an internal link to the natural order represented by *tian*. Conduct originating from this state is moral. Inner turmoil disrupts our capacity to resonate with *Tian*; conduct emerging from such inner turmoil is bound to bring disharmony and have destructive consequences. In the following quotation, Li takes these ideas and links them to the *Daode jing*, with a slight amendment:

> Lord Lao said, "[If] people are able to be constantly clear and tranquil, heaven and earth will return in their entirety [to form a unity with them[44]]."

> 人能常清靜, 天地悉皆歸.[45]

Thus, for Li, the inner state serving as the foundation for moral conduct is also the state that unifies us with sky and earth. As is observed in both the *Daode jing* and the *Neiye*, the reduction of desires and the emptying of the mind that those texts advocate become fully integrated with the Ru notion of inner balance and its function as a means for us to listen to the dictates of *Tian* (*Tianming* 天命). This listening translates into productive human conduct on the earth informed by reduced desires, clear minds, and unbiased judgments. Li brings us back to the Celestial Masters' belief that we must attend to the regulating patterns of the sky (father) so that we treat fellow human beings and the earth (mother) on which we depend for sustenance with a natural deference and respect. Of course, as an inner alchemist seeking to produce the metaphorical Golden Elixir, and like the *Mencius, Daode jing, Neiye, Taiping jing,* and the works of Wang Chongyang, Li views this attuning with *Tian* or Dao as conducive to full human health and longevity. For him as with the others, moral conduct is quite literally healthy conduct. To behave in a manner contrary to these insights is not so much wrong as simply self-destructive.

## Concluding Comments

So what can these teachers and texts contribute to a different story, one that will reorient us in such a way that we expand our horizon beyond primarily human concerns and begin to attend to the dire state of the mother and father that nurture us? Firstly, we do not find a foundation for ethics and moral conduct located in the closed domain of human social interactions. Instead, we find it in an outward turn to our wider context informed by the natural patterns and rhythms embodied in the oscillations of the sun and moon, and their associated rhythms of the seasons and tides, the growth and decline of plants, and the birth, growth, and death of animals, including human beings. We see human beings as situated not at the pinnacle of creation but instead in an uncertain intermediary position between the sky and earth. This position requires constant attentiveness to our inner lives if we are to embody conduct that conduces to harmony outwardly. Our inner lives and our health depend for their stability on actions not dictated by the whims of desire, expressed partly through often unnecessary compulsive consumption, and not conditioned by emotional turmoil associated with the restlessness of feeling we never quite have enough. At the root of this ethical dynamic is an assumption that our conduct is at its best when it is responsive to concerns and forces much greater than ourselves but that those greater forces are really not separate from us. Instead, they function as a kind of cosmic ecological system—as a wider system of fluid relationships that, if attended to, can bring that wider "ecology" and us into a state of mutually beneficial homeostasis.

The problem of personal sacrifice is transformed if we can understand that our emotional and bodily health are deeply interwoven into the fabric of our social interactions and that such action is best regulated through attentiveness to our inner lives. Further, those inner lives, made stable by a decrease in anxiety associated with desires and emotional turmoil, can open us up to awareness of our wider ecological context beyond that of the merely human. Not to attend to inner and outer dimensions of our existence is to suffer a deeper sacrifice—sacrifice of our personal well-being and that of the community of social and, more fundamentally, ecological relationships that nourish us as a mother nourishes her children.

The difficulty of how to affect such a fundamental change of perspective can feel insurmountable, but these human insights are nonetheless there for us to consider. In China as elsewhere they have not held sway, but that does not diminish their significance as cultural and intellectual resources from which we can draw. Let us hope that the impetus for reconsideration of this radically different story does not have to reach catastrophic proportions before we are willing to be moved by it. We can perhaps be reassured by the fact that the current prevailing narrative is a very recent invention. Hope lies in the very fact that our narratives are contingent and subject to continuous transformation.

## Notes

1   See also Norman J. Girardot, James Miller, and Liu Xiaogan, eds., *Daoism and Ecology: Ways within a Cosmic Landscape* (Cambridge, MA: Harvard University Press, 2001).

2   Claudia Senik, "Wealth and Happiness," *Oxford Review of Economic Policy* 30, no. 1 (2014): 92–108. Senik provides a rich review of literature on both sides of the debate and cites compelling research findings illustrating some of the strong evidence of a correlation between both wealth and income and life satisfaction and happiness, despite noting the attendant ecological ruin we are unleashing.

3   Senik, "Wealth and Happiness," 101.

4   Senik, "Wealth and Happiness," 101. This conclusion is based on findings of Betsey Stevenson and Justin Wolfers, "Economic Growth and Subjective Well-Being: Reassessing the Easterlin Paradox," *Brookings Papers on Economic Activity* 39, no. 1

(Spring 2008): 1–102, and Angus Deaton, "Income, Health and Well-being around the World: Evidence from the Gallup World Poll," *Journal of Economic Perspectives* 22 (2008): 53–72.

5 Cary Funk and Brian Kennedy, "The Politics of Climate," Pew Research Center (October 2016): 17.

6 Martin Patchen, "Public Attitudes and Behaviour about Climate Change: What Shapes Them and How to Influence Them." Purdue Climate Change Research Centre, Outreach Publication 0601 (October 2006): 5.

7 Patchen, "Public Attitudes," 6.

8 John D. Sterman and Linda Booth Sweeney, "Understanding Public Complacency about Climate Change: Adults' Mental Models of Climate Change Violate Conservation of Matter," *Climatic Change* 80 (2007): 213–38, demonstrates that even highly educated MIT graduate students appear not to comprehend the dire need for immediate action, believing that we have time to consider options that will have no economic costs.

9 Of course, one could point to continued systemic injustices. Although slavery was officially abolished in the United States, blacks constitute nearly half of the prison population in that country despite representing only 13 percent of the population. Thus, injustices have been re-inscribed through different means. See "Criminal Justice Fact Sheet," National Association for the Advancement of Colored People (NAACP), https://www.naacp.org/criminal-justice-fact-sheet/ (accessed February 1, 2017).

10 "What's in a Cigarette?" American Lung Association, https://www.lung.org/stop-smoking/smoking-facts/whats-in-a-cigarette.html (accessed February 1, 2017).

11 Samuel Perreault, *Impaired Driving in Canada*, 2015 (Ottawa: Canadian Centre for Justice Statistics/Statistics Canada, 2016), 3.

12 Alex Evans, *The Myth Gap: What Happens When Evidence and Arguments Aren't Enough* (London: Eden Project Books, 2017), ch. 1, sec. 1.

13 Evans, *The Myth Gap*, ch. 2.

14 Ronald Wright, *A Short History of Progress* (Toronto: House of Anansi Press, 2004), 129.

15 See also Paul Crowe, "Daoist Heritage Today," in *The World's Religions: Continuities and Transformations*, 2nd ed., ed. Peter Clark and Peter Beyer (London: Routledge, 2008), 122–35.

16 The earliest uses of these terms, translated as "Daoist" and "Daoism," actually antedate the Daode jing 道德經 (Scripture of the Way and Virtue) and the eponymous Zhuangzi 莊子 by several centuries. Despite this, many view these two texts as quintessentially Daoist. While they do contain many ideas foundational to later movements, and to those who today identify themselves as Daoist, they are but two of nearly fifteen hundred texts in the extant Ming Daoist Canon published in 1445.

17 A detailed account of these developments is found in Kidder Smith, "Sima Tan and the Invention of Daoism, 'Legalism,' 'Et Cetera,' " *Journal of Asian Studies* 62, no. 1 (2003): 129–56.

18 子曰: 為政以德, 譬如北辰, 居其所而眾星共之. *Lunyu* 2.1. A very similar assumption is found at *Lunyu* 12.19.

19 上德無爲而無以爲. *Daode jing*, 38.

20 In the *Neiye*, *qi* 氣 or *jing* 精 are used as synonyms. The later meaning of *jing* 精 as designating the vital force supporting reproduction is not evident in this early text.

21 Harold Roth, *Original Tao: Inward Training (Nei-yeh) and the Foundations of Taoist Mysticism* (New York: Columbia University Press, 2004), 46.

22 See *Neiye* 內業, Chinese Text Project (CTEXT), http://ctext.org/guanzi/nei-ye (accessed March 25, 2017).

23 This definition is provided in section 8 of the *Neiye*. See Roth, *Original Tao*, 60.

24 "Integrity" is how D.C. Lau defines this term in his translation of the *Mengzi*. Taken more literally, *zhi* 直 means "straight" or "direct," and by extension something close to "being forthright" or "integrity." D.C. Lau, *Mencius* (London: Penguin Classics, 2003), 2.A.2., 33.

25 Lau, *Mencius*, 2.A., 60. Chinese text: *Mengzi* 孟子, CTEXT, http://ctext.org/mengzi/gong-sun-chou-i (accessed March 26, 2017).

26 These enduring and widespread assumptions are restated and given formal expression several centuries later in the Han dynasty medical text the *Huangdi neijing suwen* 黃帝內經素聞 (Yellow Emperor's Inner Classic-Basic Questions). The best English-language translation and study of this text is Paul U. Unschuld and Hermann Tessenow in collaboration with Zheng Jinsheng, *Huang Di nei jing su wen: An Annotated Translation of Huang Di's Inner Classic—Basic Questions* (Berkeley: University of California Press, 2011).

27 Lau, *Mencius*, 4.A.15. Chinese text: *Mengzi* 孟子, CTEXT, http://ctext.org/mengzi/li-lou-i (accessed March 26, 2017).

28 Lau, *Mencius*, 7.A.21. Chinese text: *Mengzi* 孟子, CTEXT, http://ctext.org/mengzi/li-lou-i (accessed March 26, 2017).

29 A detailed discussion of this phenomenon can be found in Mark Csikszentmihalyi, *Material Virtue: Ethics and the Body in Early China* (Leiden, NL: Brill, 2004).

30 Robert G. Henricks, *Lao Tzu's Tao Te Ching: A Translation of the Startling New Documents Found at Guodian* (New York: Columbia University Press, 2000), 72.

31 Sarah Queen observes that the principal source for Dong's ideas is the *Chunqiu fanlu* 春秋繁露 (Luxuriant Dew of the Spring and Autumn Annals), though it was not composed until two or three centuries after his death. It is generally conceded, however, that this work does reflect a body of consistent insights attributable to Dong. See Queen, *From Chronicle to Canon: The Hermeneutics of the Spring and Autumn Annals According to Tung Chung-shu* (Cambridge: Cambridge University Press, 1996).

32  *Chunqiu fanlu*, ch. 56, CTEXT, http://ctext.org/chun-qiu-fan-lu/ren-fu-tian-shu (accessed April 2, 2017).

33  This idea is explained in *Chunqiu fanlu*, ch. 57, as follows: 試調琴瑟而錯之，鼓其宮，則他宮應之，鼓其商，而他商應之，五音比而自鳴，非有神，其數然也. See CTEXT, http://ctext.org/chun-qiu-fan-lu/tong-lei-xiang-dong (accessed April 2, 2017).

34  For examples, see *Chunqiu fanlu*, ch. 35, CTEXT, http://ctext.org/chun-qiu-fan-lu/shen-cha-ming-hao (accessed April 2, 2017).

35  The most detailed study accompanied by a complete translation is Barbara Hendrischke, *The Scripture of Great Peace: The Taiping jing and the Beginnings of Daoism* (Los Angeles: University of California Press, 2007).

36  "CT" refers to the text number in Kristofer Schipper, ed., *Concordance du Tao-tsang: Titres des ouvrages* (Paris: EFEO, 1975). "HY" refers to the text number in *Weng Dujian* 翁獨健. *Daozang zimu yinde* 道藏子目引得 (Combined Indexes to the Authors and Titles of Books in Two Collections of Taoist Literature), Harvard-Yenching Institute Sinological Index Series, no. 25 (Beijing: Yenching University, 1935; repr. Taipei: Chengwen, 1966).

37  *Taiping jing*, 45.1b.

38  *Taiping jing*, 45.1b–2a.

39  *Taiping jing*, 45.3a–b.

40  *Chongyang lijiao shiwu lun*, 2b.

41  See verse 15 of the *Chongyang lijiao shiwu lun*, 6a and the second in verse 7, 3b.

42  *Chongyang zhenren Danyang ershisi jue* 重阳真人授丹阳二十四訣 (Twenty-four Explanations Offered by True Man Chongyang to Dan-yang; HY1149, CT1158), 4b.

43  CTEXT, http://ctext.org/liji/zhong-yong (accessed May 30, 2017).

44  The addition in square brackets is based on Li's commentary on this phrase. *Taishang laojun shuo qingjing jingzhu* 太上老君說常清靜妙經注, CT755, HY754, 2b.

45  *Taishang laojun shuo qingjing jingzhu* 太上老君說常清靜妙經注, CT755, HY754, 2a.

# SIX

# The Ecological Virtues of Buddhism

DAVID R. LOY

## Summary

TRADITIONAL BUDDHIST VALUES PROMOTE ECOLOGICAL VIR-
tues. The earliest texts identify the causes of suffering as craving and
delusion, and encourage simplicity instead. Other early scriptures
demonstrate sensitivity to the natural world and concern for its well-being.
The Buddha's innovative understanding of karma focuses on the importance
of transforming habitual motivations and developing positive character
traits. The Mahayana tradition, which developed later, emphasizes the *bodhisattva* path, adding a more active role for compassion in how we live from
day to day. Today some Buddhists are developing the concept of an "ecosattva" path, which combines personal contemplative practice with an activism
that not only reduces one's own ecological footprint but also seeks ways to
promote the social and economic changes that are necessary for a more sustainable world.

What makes an act morally significant? There are many ways to respond to this question, but three different accounts are the most popular: consequentialism (are the results good or bad?), deontology (does it accord with moral laws?), and virtue ethics (what would a virtuous person want to do in that situation?). By no coincidence, these three types of ethical theory correspond to three distinguishable aspects of a moral action: the results that I seek, the rule I am following (or breaking), and my mental intention or motivation when I do it.

Although these aspects of moral consideration cannot be completely separated from each other, we can emphasize one more than the others. Insofar as Buddhism offers a path that leads to awakening, Buddhist morality is consequentialist. Rule-based behaviour is also important, such as the five basic precepts that all Buddhists are expected to follow, and the 227 Vinaya rules that regulate the lives of male monastics (female monastics follow more rules). Yet each of the five precepts is a "mindfulness training": not a commandment imposed by someone or something outside, but a *vow to oneself* that involves cultivating a particular trait or virtue. The many rules that regulate the life of a Buddhist monastic can be understood in a similar fashion. When problems arose within the monastic order and were brought to the attention of the Buddha, he would make a ruling about the best way to act in such situations, in order to avoid "hindrances" (*nivaranani*) that might interfere with one's spiritual development. Not the letter but the spirit of the rules was what counted. According to the *Mahaparinibbana Sutta*, at the end of his life the Buddha even said that many of the "minor" rules could be abandoned.

Instead of focusing on the duality of good versus evil, however, Buddhism characterizes actions as "wholesome/unwholesome" according to their likely results, including their effects on one's habitual ways of thinking and acting. In short, Buddhist teachings regarding morality are mostly a version of virtue ethics: the point is to develop certain character traits. And those traits are especially important today, because they are the kinds of virtues that are needed in our time of ecological crisis.

## The Five Precepts

The five basic precepts involve abstaining from five unwholesome types of acts: killing living beings, "taking what is not given," misconduct concerning sense-pleasures, "false speech," and "alcoholic drink or drugs that are an opportunity for heedlessness." From an ecological perspective the first two "mindfulness trainings" stand out: not to kill any living being and not to take something that has not been given to me. Notice that the first precept is broader than the injunction in the Mosaic Decalogue against killing another human being: other species are also not to be harmed. The second precept is usually understood to forbid stealing, but again there are broader environmental implications.

These implications are embodied in early Buddhist texts (in the Pali Canon) that reveal sensitivity to the beauties of nature and respect for its various beings. A good example is the Jataka tales that purport to describe the previous lives of the Buddha before he became the Buddha. In many of those fables he is born as an animal, and in some of the best-known ones the future Buddha sacrifices himself for "lower animals," such as offering his rabbit body to a weak tigress so that she can feed her starving cubs. Such tales imply that the welfare of every living being is deserving of our concern. Notably, all beings in the Jataka stories are able to feel compassion for others and act selflessly to help ease their suffering. In contrast to "survival of the fittest," these fables offer a vision of life in which all species remain deeply interconnected, and therefore also responsible for each other.

This compassion is not limited to the animal realm. According to the traditional biographical accounts of his life, the Buddha was born under trees, meditated under trees, experienced his great awakening under trees, often taught under trees, and died beneath trees. Unsurprisingly, he often expressed his appreciation for trees and other plants. In one text, a tree spirit appears to the Buddha in a dream, complaining that it had been chopped down by a monk. The next morning the Buddha prohibited *sangha* members from cutting down trees. Monastics are still forbidden from cutting off tree limbs, picking flowers, even plucking green leaves off plants.

Those precise rules do not apply to Buddhist lay people, but it is not difficult to extend the general principles involved to include prohibition of actions and policies that harm ecosystems. Whether or not the extinction of a particular species (rather than a particular individual) violates the letter of the first precept, the species extinction event that is happening today certainly violates its spirit. And "not taking what is not given," the second precept, seems incompatible with human exploitation and abuse of the biosphere.

But why is it important to follow these precepts, according to Buddhism? The traditional answer is that when I violate them, the person I hurt most of all is myself—that is, I aggravate my own *dukkha* ("suffering" in the broad sense: dissatisfaction, stress, dis-ease). That brings us to the Buddhist account of karma, which, although often understood in a consequentialist way, encourages the development of wholesome character traits.

## Karma

Most of the Buddha's contemporaries believed in karma, but they understood it mechanically and ritualistically. According to Brahmanical teachings, performing a sacrifice in the proper fashion would lead to the desired consequences. One of the Buddha's most important innovations is the way he transformed this formulaic approach into a moral principle by focusing on *cetana* (motivations, intentions). For example, the *Dhammapada*, probably the most popular early Buddhist text, begins by emphasizing the pre-eminent importance of our mental attitude:

> Experiences are preceded by mind, led by mind, and produced by mind. If one speaks or acts with an impure mind, suffering follows even as the cart-wheel follows the hoof of the ox.
>
> Experiences are preceded by mind, led by mind, and produced by mind. If one speaks or acts with a pure mind, happiness follows like a shadow that never departs.[1]

As commonly used, the term "karma" (*kamma* in Pali) refers to the results of something done in the past: for example, when something happens to me, I may respond: "Well, that must be my karma." However, the original term literally means "action." Buddhism has other words for the consequences of an action: *vipaka* is the "result," also known as *phala*, the "fruit." As this suggests, the true focus of the Buddhist approach is not on the consequences (effects) of one's actions (causes) but on the actions themselves. In most popular understandings, the "law" of karma and rebirth offers a way to get a handle on how the world will treat us in the future. This also implies, more immediately, that we should accept our own causal responsibility for whatever is happening to us now, as a consequence of what we must have done earlier.

However, this fatalistic perspective puts the cart before the horse, because it misunderstands the revolutionary significance of the Buddha's reinterpretation. His emphasis on *cetana* makes karma the key to one's spiritual development: *how my life-situation can be transformed by transforming the motivations of my actions right now*. The basic point is that karma is not something the self *has*; instead, it is what the self *is*, because one's sense of self is transformed by one's conscious choices. Just as my body is composed of the food eaten and digested, so my character is composed of those choices, which when repeated become habitual tendencies, whether wholesome (virtues) or unwholesome (character defects). In this way "I" am (re)constructed by my consistent, recurring mental attitudes. By choosing to change what motivates me—what motivations I act upon—I change the kind of person I am.

From this perspective, we experience the consequences not just for what we have *done* but most of all for what we have *become*, and what we intentionally do is what makes us who we are. An anonymous verse expresses this well:

Sow a thought and reap a deed
Sow a deed and reap a habit
Sow a habit and reap a character
Sow a character and reap a destiny

(Of course, it's not quite so simple: remember the fate of Jews in Nazi Germany. This reminds us that individual transformation by itself is not enough. We also need to be concerned about social transformation—that is, about social justice. What happens to us is not separate from what happens to other people.)

What kind of thoughts (motivations) do we need to sow? Buddhism traces back our *dukkha* to the "three poisons" of greed, ill will, and delusion. These problematic motivations need to be transformed into their positive counterparts: generosity, loving-kindness, and the wisdom that realizes our interdependence with others.

This perspective helps us to understand the important role of confession and repentance. They are our way of acknowledging, both to others and to ourselves, that we are striving not to allow something deplorable that we have done to become (or remain) a habitual tendency.

Note that such an understanding of karma does not necessarily involve another life before or after this one—although acting according to it would also be the best way to prepare for one's next lifetime (if there is rebirth). As Spinoza expressed it at the end of his *Ethics*: "Blessedness [*beatitudio*] is not the reward of virtue, but virtue itself."[2] Looked at from the other side, I am punished not *for* my unwholesome actions but *by* them. Whether my habits are wholesome or unwholesome, I become the kind of person who does that sort of thing.

To become that kind of person is also to live in that kind of world. That is because when one's motivations change, the world changes: when we relate differently to the world, the world tends to relate differently to us. Insofar as we are interdependent with other people, our ways of acting in it involve feedback systems that incorporate them. People not only notice what we do, they notice why we do it. I may fool people sometimes, yet over time my character becomes revealed as my intentions become evident. The more I am motivated by greed, ill will, and delusion, the more I must manipulate others to get what I want, and consequently the more alienated I feel (and the more alienated from me others feel) when they realize that they have been manipulated. This mutual distrust encourages both sides to manipulate more.

On the other side, the more my actions are motivated by generosity, loving-kindness, and the wisdom that recognizes our interdependence, the more I can relax and open up to the world. The more I feel part of the world and genuinely connected with others, the less I will be inclined to take advantage of them, and consequently the more inclined they will be to trust and open up to me. In such ways, transforming my own motivations not only transforms my own life; it also affects those around me, since what I am is not separate from what they are.

What are the ecological implications of this perspective? Buddhist teachings about karma have traditionally been understood in individual terms: my karma is different from yours. Recently, however, the issue of *collective karma* has been discussed, usually focusing on consequences: for example, might the group effects of a natural disaster such as an earthquake be the result of improper behaviour by that group? But perhaps that question, too, puts the cart before the horse: what about collective karma understood as *collective motivations*? The more I feel separate from others—that my well-being is separate from theirs—the more likely I am to act in a way that is indifferent to their well-being. Does the same hold true communally? Today the ecological consequences of that attitude are hard to avoid: our collective sense of separation from the earth rationalizes our collective exploitation of its ecosystems, which are often understood as nothing more than resources for us to utilize.

The problem with this, as we are discovering, is that when we damage nature, we are damaging ourselves. Is the ecological crisis, then, an obvious example—maybe even the "best" possible example—of collective karma?

## Poverty

The ecological virtues of Buddhism extend beyond its understanding of karma. We also need to consider Buddhists attitudes toward wealth/poverty, which have important implications for how we relate to the earth now, as well as how we may need to adjust to future eco-constraints.

According to Buddhism, poverty is bad insofar as it involves *dukkha*. The point of the Buddhist path is to end one's *dukkha*, and that does not imply we must distinguish between worldly *dukkha* and some other "spiritual" sort. Buddhism values non-attachment to material goods, and promotes the virtue of having fewer desires, yet that is not the same as encouraging physical poverty.

Poverty, as understood in early Buddhism, consists in lacking the basic material requirements for leading a decent life free from hunger, exposure, and disease. Buddhism recognizes the importance of such minimum needs even in the case of those who aspire to its spiritual goals, and in fact the basic needs of a monk or nun provide a useful benchmark for measuring that level of subsistence below which any human being should not be allowed to fall. The traditional "four requisites" of a Buddhist renunciant are: food sufficient to alleviate hunger and maintain one's health; clothing sufficient to be socially decent and to protect the body; shelter sufficient for serious engagement with cultivating the mind; and health care sufficient to cure and prevent disease. People who voluntarily renounce worldly possessions and pleasures in favour of a life of such minimal needs are viewed as belonging to the community of "noble ones" (*ariyapuggala*). For a more comprehensive evaluation of deprivation, however, it is necessary to take into account the moral quality of people's lives.

So Buddhism draws attention to the fact that the single-minded pursuit of material wealth cannot make human beings happy. The Buddha spoke of the four kinds of happiness (*sukha*) attained by householders: possessing enough material resources, enjoying those resources, sharing them with relations and friends, and not being in debt. More important than any of them, he emphasizes, is the happiness of leading a blameless life. Elsewhere, the Buddha teaches that the greatest wealth is contentment. A world in which envy (*issa*) and miserliness (*macchariya*) predominate cannot be considered one in which poverty has been eliminated. According to the second noble truth of the Buddha, the cause of *dukkha* is *tanha* (craving). Whenever human beings gain an intense acquisitive drive for some object, that object becomes a cause of suffering. Such objects are compared to the flame of a torch carried against the wind, or to a burning pit of embers: they involve much anxiety but very little satisfaction—an obvious truth often overlooked by immediately turning our attention to the next craved object. For Buddhism such a proliferation of wants is a basic cause of unnecessary suffering.

Poverty can never be overcome by proliferating more desires that are to be satisfied by consuming more goods and services. In some places this may result in the elimination of material poverty for some or even many people, yet at the cost of promoting a different kind of poverty, with other types of *dukkha* that are even more harmful. In short, there is a fundamental poverty built into a consumer society.

Unless they have been seduced by the utopian dream of a technological cornucopia, it would never occur to most "poor" Indigenous people to become fixated on fantasies about all the things they might have. That is because their ends are an expression of the means available to them. It is presumptuous to assume that they must be unhappy and that the only way to become happy is to start on the treadmill of a lifestyle dependent on the market and increasingly preoccupied with consumption.

All this is better conveyed with a Tibetan Buddhist analogy. The world is full of thorns and sharp stones (and now broken glass, as well). What should we do about this? We can try to pave over the entire earth, or we can wear shoes. "Paving the whole planet" is a good metaphor for how our collective technological and economic project is attempting to make us happy. The other solution is to learn how to make and wear "shoes," so that our collective ends become an expression of the renewable means that the biosphere provides.

THE ECOLOGICAL VIRTUES OF BUDDHISM  ·  125

The fundamental human problem is not the technological and economic issue of meeting all our material "needs"—something ecologically impossible—but the psychological and spiritual need to understand and control our own minds: that is, learning to wear shoes.

From a Buddhist perspective, then, the common assumption that "income/consumption poverty" is the same as ill-being is a mistake. That becomes especially important when we realize that the world's growing population (7.8 billion people, as of this writing) cannot all attain the kind of consumerist lifestyle that so many modern "economized" people take for granted—or aspire to.

That brings us to the heart of the matter. I wonder if, for many people today, material well-being has become increasingly important because of a widespread loss of faith in any other possibility of fulfillment. Perhaps increasing our "standard of living" has become such a compulsive pursuit for us because it substitutes for traditional religious values that many people can no longer believe in, or, more precisely, because consumerism has actually become a kind of secular religion for us. The ecological crisis, with its implicit challenge to a consumerist lifestyle, signifies the end of this collective fantasy.

## Mahayana

Everything discussed so far considers basic principles that apply more or less to all Buddhist traditions. By the first century BCE, however, a new type of Buddhism had begun to emerge in India, which emphasized certain virtues that have other ecological implications today.

The Mahayana tradition highlighted compassion (*karuna*) as the primary character trait to be developed. Its perspective on moral virtue involved "a dynamic, other-regarding quality, rather than primarily concerned with personal development and self-control."[3] Instead of focusing on one's own awakening, as the solution to one's own *dukkha*, adherents were encouraged to expand the sphere of their concern in order to address the *dukkha* and the awakening of others—in fact, the awakening of *everyone*. "We should have this [compassion] from the depths of our heart, as if it were nailed there. Such compassion is not merely concerned with a few sentient beings such as friends and relatives, but extends up to the limits of the cosmos, in all directions and toward all beings throughout space."[4]

Two oft-quoted verses from the *Bodhicaryavatara*, by the eighth-century Indian monk Santideva, exemplify this attitude:

> May I be the doctor and the medicine
> And may I be the nurse
> For all sick beings in the world
> Until everyone is healed [3:8]

> All the misery in the world
> derives from desiring happiness for oneself;
> all happiness in the world arises from
> desiring the happiness of others [8:129][5]

This emphasis on compassion was not completely new. According to the *Karaniya Metta Sutta*, Shakyamuni Buddha recommended suffusing the whole world with benevolence:

> Whatever living beings there be—feeble or strong, long, stout, or medium, short, small, or large, seen or unseen, dwelling far or near...may all beings, without exception, be happy!...Just as a mother would protect her only child even at the risk of her own life, even so, let one cultivate a boundless heart toward all beings. Let one's thoughts of boundless love pervade the whole world—above, below, and across without any obstruction, without any hatred, without any enmity.[6]

Shakyamuni also encouraged practitioners to develop and reside in what came to be known as the four "divine abodes" (*brahmaviharas*): *metta* (loving-kindness); *karuna* (compassion for the suffering of others); *mudita* (empathetic joy in the happiness of others); and *upekkha* (imperturbable equanimity).

Mahayana, however, expanded the role of compassion by offering a new articulation of what it means to be a Buddhist practitioner: the *bodhisattva path*. Whereas the five precepts of early Buddhism focused on what not to do—don't kill living beings, don't take what is not given, etc.—the career of the *bodhisattva* emphasizes character traits that should be developed to the highest possible degree.

In the early texts preserved in the Pali Canon, the term *bodhisattva* was used to refer to the earlier lives of Shakyamuni before he became the Buddha.

According to a common sectarian account, there was a conspicuous difference between his accomplishment and that of his followers, the *arahants* who awakened later by following his teaching. By definition, an *arahant* (literally "one who is worthy") has achieved the same *nibbana* as the Buddha, yet the Buddha was nonetheless special because he devoted himself wholeheartedly to helping everyone awaken. This led to the development of a more altruistic model of how to live. Today, the *bodhisattva* path is increasingly understood by contemporary Buddhists as a spiritual archetype that offers a new vision of human possibility—an alternative to rampant self-preoccupied individualism, including any approach to Buddhist practice that is concerned only about one's own personal awakening.

Thus the *bodhisattva* path begins by arousing the *bodhicitta*, which is the desire and determination to awaken in order to promote the well-being of all beings, not just one's own. Thereafter, one focuses on developing the six *paramitas* (perfections), which all involve attitudes and actions that are cultivated and performed in a non-egocentric way.

The first, *dana* (literally, "giving") is sometimes said to contain all the other five. It involves generosity to other individuals or institutions (such as a Buddhist monastic, or the *sangha* community generally) with no expectation of any return or reward.

*Sila*, which can be translated as "virtue," "right conduct," or "discipline," incorporates the ethical precepts of early Buddhism. The emphasis is not on obedience or obligation, but developing self-restraint and greater awareness of the effects of one's actions.

*Ksanti* (patience) means an endurance that never takes offence or avoids an uncomfortable situation. The *Dhammapada* describes it as the "foremost austerity." In an early text, the Buddha exhorts his followers not to become hateful or speak angrily even if one's limbs were being sawed off by bandits.

*Virya* is variously translated as "energy," "enthusiasm," or "sustained effort." It involves extreme perseverance or diligence: never giving up, in order to avoid what is unskillful and to accomplish what is wholesome.

*Dhyana* (meditation) refers to the cultivation of mental concentration or contemplative practices, which are considered necessary in order to awaken.

*Prajna* (literally "highest knowing") is the wisdom that comes with awakening; according to the Mahayana understanding this is the realization that everything is "empty" of self-being, including oneself.

Needless to say, the four "divine abodes" and these six perfections involve the development of character traits that are increasingly important for our individual and collective responses to the environmental crisis. They have traditionally been understood in personal terms, as applicable to my own spiritual development and assisting that of others, because helping others on their spiritual path turns out to be an important part of my own path. Practising the six perfections, I devote myself to helping others; and by helping others I "perfect" (develop further) my perfection-practices.

Today, however, social justice issues and especially the ecological crisis are prompting some contemporary Buddhists to reconsider and expand their understanding of the *bodhisattva*. How might customary conceptions of that path be adapted to make Buddhist teachings more relevant to what is not only the great challenge of our time but also perhaps the greatest challenge that humanity has ever faced?

## The Ecosattva Path

The virtues developed on the *bodhisattva* path, and encouraged by the Buddhist tradition generally, support adaptation to and mitigation of new ecological constraints. But are they a sufficient response? Insofar as the ecological crisis may be symptomatic of a deeper problem—an effect, perhaps, of some other cause—it is important to investigate what the root of that crisis might be. Some diseases can only be treated symptomatically, but whenever possible it is preferable to address the source as well. Does Buddhism have something else to offer here that can contribute to that inquiry? And does this imply a new direction for *bodhisattva* activity?

The earlier discussion of karma mentioned the "three poisons" of greed, ill will, and delusion. Buddhist teachings do not say much about evil per se, but these three poisons are also sometimes described as the three roots of evil, because they are the closest Buddhism comes to offering an explanation for our dissatisfaction: when what I do is motivated by any or all of them (and the three tend to reinforce each other, of course), my actions are unwholesome and tend to result in *dukkha*.

From a Buddhist perspective, then, it is quite significant that the three poisons can also function collectively. Today we not only have much more powerful technologies than in the Buddha's time, we also have more powerful institutions, which operate according to their own logic and motivations—in effect, then, *they take on a life of their own.* The Buddhist emphasis on motivation can therefore provide a distinctive perspective on some of our most important social structures. Arguably, the ecological crisis can be traced back, in part, to the fact that our present *economic system* institutionalizes greed, our *militarism* institutionalizes ill will, and the *corporate media* institutionalize a delusion that supports the other two.

If *greed* is defined as "never having enough," that also applies collectively: consumers never consume enough, corporations are never profitable enough, their share value is never high enough, our GNP is never large enough... In fact, we cannot imagine what "big enough" might be. Our economic system is constructed in such a way that it must keep growing, or else it tends to collapse. But why is *more* always better if it can never be *enough*?

Consider the stock market, high temple of the economic process. On one side are many millions of investors, most anonymous and largely unconcerned about the details of the companies they invest in, except for their profitability and share prices. In many cases investors do not know where their money is invested, thanks to mutual funds. Such activity is not evil, of course: successful investors are highly respected, even idolized (Warren Buffett is "the sage of Omaha"). On the other side of the market, however, the desires and expectations of those millions of investors become transformed into an impersonal and unremitting pressure for growth and increased profitability that every CEO must respond to, and preferably in the short run.

Consider, for example, the awkward situation for the CEO of any large corporation who suddenly awakens to the dangers of climate change and wants to do everything he (it's usually a *he*) can to address this challenge. If what he tries to do threatens corporate profits, he is likely to lose his job. And if that is true for the CEO, how much more true it is for everyone else down the corporate hierarchy. Corporations are legally chartered so that their first responsibility is not to their employees or customers, nor to the members of the societies they operate within, and certainly not to the ecosystems of the earth, but to their stockholders, who with few exceptions are focused primarily on return on investment.

Who is responsible for this collective fixation on growth? The point is that this *system* has its own in-built motivations, quite apart from the motivations of the employees, including administrators and executive board members, who will be replaced if they do not serve that institutional motivation. And all of us participate in this process in one way or another, as employees, consumers, investors, pensioners, and so forth, although with very little (if any) sense of personal responsibility for the collective result. Any awareness of what is actually happening tends to be diffused in the impersonal anonymity of this economic process. We are simply doing our jobs.

The foremost example of *institutionalized ill will*, the second "root of evil," is militarism. This is ecologically important, given the role of military activity in contributing directly to environmental degradation; the US military, for example, is the largest consumer of oil on the planet. Measured by the power of its armed forces and the resources devoted to them—well over half a trillion dollars every year—the United States is the most militarized society in world history. The need to "defend ourselves" apparently requires well over 700 military installations overseas, and more than 900 at home. No wonder there is so little money left over for education, health, and social services.

To justify that colossal expense, the military needs an enemy, of course. The end of the Cold War therefore created a problem for the Pentagon, but terrorism has replaced the preoccupation with communism. The so-called War on Terror is already by far the longest war in US history, and it may never come to an end. Using drones to assassinate suspected terrorists, along with anyone else who happens to be nearby, ensures a dependable supply of angry people who have good reason to hate the United States. If terrorism is the war of the poor and disempowered, then war is the terrorism of the rich.

Much more could be said about the ways that militaristic attitudes are intertwined with lucrative arms research, manufacturing, and international sales— together supporting collective *dukkha* on a vast scale—but let us turn instead to the third poison, *institutionalized delusion*.

Genuine democracy requires an independent and activist press, to expose abuse and discuss important political issues. As they have merged and consolidated, however, the major US media (six mega-corporations now control 90 percent of the market, down from fifty companies in 1983) have abandoned all but the pretense of objectivity. Since they are profit-making institutions whose bottom line depends upon advertising revenue, it is not in their own best interest to question the grip of consumerism, a value system profoundly incompatible with Buddhist virtues, as we have seen.

An important part of genuine education is realizing that many of the things we thought were natural and inevitable (and therefore should be accepted) are in fact conditioned (and therefore can be changed). The world does not need to be the way it is; there are other possibilities. The present role of the media is to foreclose most of those possibilities by confining public awareness and discussion within narrow limits. With few exceptions, the world's developed (or "economized") societies are now dominated by a power elite that moves seamlessly from corporate boardrooms to top government positions and back again, because there is little if any difference in their world view or goals: never-ending economic expansion. The media "normalizes" this situation, so that we accept it and continue to perform our required roles, especially the frenzied production and consumption necessary to keep the economy growing.

If the Buddha was correct that greed, ill will, and delusion are the "three roots of evil"—the basic motivations that cause suffering—and if they have indeed become institutionalized, the environmental implications are matters for deep and urgent concern. The connection of such an interlocking economic-military-media-and-political structure with the incessant degradation of the natural world is hard to miss. We should not be surprised that this self-enclosed system seems unable to address the ecological crisis sufficiently: its basic project, predicated on continuous growth, is incompatible with the well-being of the biosphere. What does this mean for the way that Buddhist values and virtues should be understood and practised now, in particular, the *bodhisattva* path? Doesn't awakening to the nature of these three institutional poisons become just as important as the individual awakening that Buddhism traditionally emphasizes?

It is difficult to avoid the implication that today the *bodhisattva* path needs to expand beyond the conventional vision, to include an activism that is concerned about not only individual transformation but also social transformation. Given the nature of Buddhist teachings, such activism will have some distinctive characteristics. Buddhist emphasis on interdependence and delusion implies not only non-violence but a politics based on compassion rather than anger. As I have argued, the basic problem is not rich and powerful "bad people" but institutionalized structures of collective greed, aggression, and delusion, which condition us to think and act in problematic ways. In addition to the individual transformation that Buddhism has traditionally emphasized, those institutions also need to be transformed. The Buddha's pragmatism and non-dogmatism—he taught, for example, that his teachings are a raft to help us awaken, not dogma to "carry on our backs"—can help to cut through the ideological quarrels that have weakened so many progressive groups. And the Mahayana emphasis on *upaya-kausalya* (skill in means)—the ability to adapt and respond successfully to new situations—foregrounds the importance of creative imagination, a necessary attribute if we are to construct a healthier way of living together on this earth, and work out a way to get there.

Acknowledging the importance of social engagement is a big step for many Buddhists, who are used to focusing on their own personal peace of mind. On the other side, those committed to social action tend to suffer from frustration, anger, depression, and burnout. The *bodhisattva* path offers what each needs, because it involves a double orientation, inner and outer, that enables earnest engagement in goal-directed behaviour without the usual attachment to results. On the one hand, the *bodhisattva* continues with his or her individual practice, which normally includes some form of daily meditation; this cultivates a basic ground of serenity that becomes especially important in especially difficult times, when one becomes overwhelmed by the magnitude of the task.

In Japanese Zen temples, practitioners daily recite the four "*bodhisattva* vows." The first one is to help all living beings awaken: "Sentient beings are numberless; I vow to liberate them all." Given that this is an unachievable task, the vow really involves reorienting the meaning of one's life, from our usual self-preoccupation to primary concern for the well-being of everyone. On a day-to-day level, what is important is not the unattainable goal but the direction of one's efforts. My point is that this is the same spirit with which eco-sattvas today need to engage with their tasks. Someone who has volunteered for a job that is literally impossible—to help every being in the universe wake up—is not going to be intimidated by present crises, no matter how hopeless they may sometimes appear.

This means that the *bodhisattva*'s activism—the *outer practice*—involves doing the best one can, without knowing what the consequences will be. Have we already passed ecological tipping points and civilization as we know it is doomed? We don't know, yet rather than being overwhelmed, *bodhisattvas* embrace "don't-know mind" (a phrase emphasized in Zen), because Buddhist practice opens us up to the awesome mystery of this impermanent world where everything is changing whether or not we notice. I grew up in a world defined by a "cold war" between the United States and the Soviet Union, which

THE ECOLOGICAL VIRTUES OF BUDDHISM  ·  133

everyone took for granted—until communism suddenly collapsed overnight. The same thing occurred with South African apartheid. If we don't really know what is happening, do we really know what is possible, until we try?

The equanimity of the *bodhisattva*-activist is due to non-attachment to the fruits of one's action, which is not detachment from the state of the world or the fate of the earth, but the ability to bounce back from disappointments, again and again. The source of that non-attachment is in the *bodhisattva's inner practice*. Meditation cultivates awareness of that equanimous dimension where there is nothing to gain or lose, no getting better or worse; this follows from the realization that nothing has any self-existence, which is *prajna* wisdom, the sixth "perfection" that *bodhisattvas* develop. The danger, for traditional Buddhist practitioners, is becoming attached to that insight and therefore becoming indifferent to what is happening in the world. The problem, for activists, is on the other side: being unaware of meditation and other forms of inner cultivation leaves them without an imperturbable ground or stable basis for their life work, and therefore tends to weaken what they are able to contribute—for example, the well-known problem of burnout.

Each by itself is incomplete; the ecosattva path has two sides, like a coin. The environmental catastrophe that has begun requires Buddhist practitioners to become engaged in the long-term process of social and ecological transformation. Those committed to activism need the patience, perseverance, serenity, and insight that the *bodhisattva* cultivates. This expanded conception of the spiritual path—which is also an ecological path—may be the most important contribution of Buddhism to our situation today.

## Ecodharma

Unsurprisingly, the teachings of traditional Asian Buddhism do not discuss climate change or the ecological crisis, but, as I have tried to show, its teachings and practices have important implications for how we understand and respond to our urgent situation. True to its emphasis on impermanence and insubstantiality, as it spread throughout Asia and now the rest of the world, Buddhism has been very adaptable, and today we see the rapid development of a new focus: *ecodharma*. For example, One Earth Sangha, based in Washington, DC, co-sponsored the first Eco-Dharma Conference at Wonderwell Mountain

Refuge in New Hampshire in 2014, and more recently has been offering online ecosattva training. The first Ecodharma Centre, located in the Spanish Pyrenees, has been offering retreats and courses since 2008. Most recently, I am pleased to be one of the founders of the Rocky Mountain Ecodharma Retreat Center, near Boulder, Colorado, which began offering retreats and workshops in 2017. Ecodharma seems to be an idea whose time has come.[7]

## Notes

1   Verses 1–2. See Gil Fronsdal, trans., *The Dhammapada* (Boston: Shambhala, 2005), 1.

2   Baruch Spinoza, *Ethics*, Book 5, proposition 42. See, for example, https:// en.wikisource.org/wiki/Ethics_(Spinoza)/Part_5. Spinoza's use of *beatitudio* has also been translated as "freedom of mind," "peace of mind," "salvation," and "happiness."

3   Damien Keown, *The Nature of Buddhist Ethics* (New York: St. Martin's Press, 1992), 131.

4   His Holiness the XIV Dalai Lama, *Aryasura's Aspiration with Commentary by the Dalai Lama and a Meditation on Compassion* (Dharamsala, India: Library of Tibetan Works and Archives, 1978), 111.

5   Stephen Batchelor, *A Guide to the Bodhisattva's Way of Life* (Glen Spey, NY: Tharpa, 2002).

6   Sutta Nipata 1.8.

7   See https://oneearthsangha.org/; https://thebuddhistcentre.com/tags/ ecodharma; https://rockymountainecodharmaretreatcenter.org/; and http:// www.ecobuddhism.org.

# SEVEN

# Never Weary of Gazing

## Contemplative Practice and the Cultivation of Ecological Virtue

DOUGLAS E. CHRISTIE

STONE BY STONE IT RISES, THIS LITTLE HOUSE BY THE SEA. SOFT, damp sand for mortar, moss for the garden, little pieces of driftwood for the roof. A tiny stone wall made from pebbles encircles the yard; a path constructed from bits of glistening seaweed winds toward the front door. I am on my belly, my young daughter beside me, working to bring this little dwelling into being. A breeze from the Pacific Ocean cools us. We began building this miniature house on a whim on this black sand beach along the Lost Coast in Northern California; but now we are going at it in earnest. We want to make it strong and beautiful. Pausing from time to time to consider our creation, we talk and laugh and exchange stories. We imagine the lives of the inhabitants of this place, what they do all day, what they care about. Gradually, a whole cosmos comes into being.

This was many years ago. My daughter is grown now, and that little house, surely, is no longer. The tides and wind would have brought it down long ago, disassembling its lovingly gathered elements, carrying them off into the ocean and returning them eventually to that very place. Or perhaps to some other place miles to the south. Or to a place unknown to any of us. Yes, it was many years ago. But that moment lives within me like a dream.

The place itself is dream-like: known as the Sinkyone Wilderness (after its original Indigenous inhabitants), it is a stretch of land in Northern California so far removed from the great metropolises of Los Angeles and San Francisco that it is difficult sometimes to believe that it is even part of the same world. Stand long enough on the vast, empty black sand beaches of the Sinkyone—amidst the presence of cormorants, terns, osprey, pelicans, and grey whales—and you may well find it hard to remember that those other places even exist. That sense of being immersed for a time in a world whole unto itself—whole and vibrant and complex and beautiful—and of losing yourself in that world, arrives as a precious gift. It is a moment almost out of time in which you behold, as if anew, the world itself. A moment of original innocence.[1]

I cherish the memory of this moment. It is emblematic of something that beckons to me continuously but so often eludes me: a simple awareness and appreciation of myself, alive in the vibrant world. Still, I recognize that the awareness I am being called to is more complex and ambiguous than this and that this place is also marked by pain, loss, and death. I remember the moment I first became aware of this. I was sitting outside the old ranch house above Needle Rock, tending a fire. Some friends had come to spend the evening and share a meal with me. Conversation drifted this way and that, and then, suddenly, one of my friends began telling the story of Sally Bell and the Needle Rock Massacre, which had occurred close to this very spot over one hundred and fifty years earlier. I listened, amazed and horrified, as he recounted the tale. Sometime later, I managed to track down Sally Bell's own account of what happened that day:

> My grandfather and all of my family—my mother, my father, and we—we were around the house and not hurting anyone. Soon, about ten o'clock in the morning, some white men came. They killed my grandfather and my mother and my father. I saw them do it. I was a big girl at the time. Then they killed my baby sister and cut her heart out and threw it in the brush where I ran and hid....
>
> Then I ran into the woods and hid there for a long time. I lived there a long time with a few other people who had got away.... After a long time,

maybe two, three months, I don't know just how long, but sometime in the summer, my brother found me and took me to some white folks who kept me until I was grown and married.[2]

The story of Sally Bell does indeed haunt the Sinkyone Wilderness. Still, it remains mostly unknown to those who visit this place. It was unknown to me until that evening sitting around the fire with friends. Now I carry it within me: an awareness of the violence that shaped this place and which is inextricably bound up with its history and ecology. There is an ancient redwood grove that now bears Sally Bell's name not far from the place where I spent the day with my daughter building our little house: it is a potent symbol and reminder of the dignity of the people who first inhabited this place, of the enduring beauty and tenacity of the life of this place, and of all we have lost through our greed, violence, and blindness. Here, as in countless other places we inhabit, beauty and violence are bound up together. There is a complex ecology here, one that includes not only the flora and fauna endemic to this part of the Northern California coast but also its long, ambiguous human history. Gradually, we are coming to realize that our relationship with such places must somehow incorporate an awareness of this entire complex reality.

Still, what kind of awareness is this? What will it mean to cultivate it? And how can the cultivation of such awareness help us to deepen the practice of what is being described in this book as ecological virtue?[3] These are the questions I propose to address in this chapter.

Above all I want to ask: How can we learn to see the world more deeply, more honestly? Also: How can we learn to acknowledge our near-habitual blindness, with all its destructive consequences, and learn to behold the world as worthy of our deepest regard and respect? In what follows, I want to suggest that the question of seeing is critical to the work of renewing our ethical relationship to the living world.

And I propose that traditions of contemplative practice can offer us important guidance on what it means to see the living world deeply and fully and how to take responsibility for both what we see and what we fail to see. What is sometimes referred to as "the contemplative gaze"—a way of seeing that is both wholehearted and all-encompassing—offers us a vision of what it means to behold the deepest, truest things in existence, including the created world, as an ever-unfolding manifestation of the divine, as a living sacrament.[4] But the contemplative gaze also invites a serious moral reckoning with the damage we have inflicted and continue to inflict upon the living world, and a willingness to accept and seek redress for all the ways we fail to see and take seriously the lives of others. In this chapter, I address these concerns from the perspective of my own Catholic faith tradition, drawing especially upon the contemplative practices cultivated by men and women across the centuries. But I do so with a full recognition that such practices resonate deeply with those found in other spiritual traditions as well as among many who do not identify with any particular spiritual tradition. The challenge of paying attention, of learning to be awake, has long had a cherished place within traditions of contemplative practice. But increasingly these concerns are being seen as common to all people of good will who seek to live responsibly in relation to the living world.

The vision of the world as sacramental—everything that exists bearing within itself a trace of the divine—corresponds to something Pope Francis describes in his 2014 encyclical *Laudato Si'* as an "integral ecology," a way of understanding ecology that refuses to separate our obligation toward the natural world from our obligation toward our fellow human beings. In terms of what we refer to so often as the "environmental crisis," Francis notes: "We are faced, not with two separate crises, one environmental and the other social, but rather with one complex crisis, which is both social and environmental."[5] So, too, with the challenge of learning to see: we are being asked to cultivate a deeper, more encompassing contemplative practice that leads to a greater awareness of both the sacramental beauty that is unfolding before us continuously as well as the loss and fragmentation that is such a prominent part of our experience of the world. This way of seeing is essential to the work of healing.

## Painfully Aware

I want to return for a moment to *Laudato Si'*, which I believe offers important insight into these very questions, especially concerning the relationship between transformed awareness and ecological virtue. It is no accident that Pope Francis begins his great encyclical with the witness of his namesake from Assisi (1181–1226). The joyous, ecstatic response to the living world for which St. Francis is renowned offers us an important model for rethinking our own relationship to the living world. The sense of awe and wonder that characterized the saint's attitude toward all living things—the sun, the moon, the air, fire, water, birds, wolves, the entire living cosmos—can and must also be ours. This is one of the clear messages of the encyclical. Yet we underestimate, I think, the challenge that this vision presents to us. St. Francis is so much more than a friendly backyard saint who loved animals; he is a strange, unpredictable, and wild figure, someone who felt the living world coursing through his veins and understood the cost of this all-involving relationship with the living world. Yes, he saw and felt the presence of other living creatures as manifestations of God's own presence in the world. Francis dwelt in a paradisal world in which the capacity to communicate with and be touched by the presence of other living beings had been fully restored to us. Think of Giotto's great, light-infused rendering of Francis's joyful sermon to the birds. Or Francis's lyrical *Canticle to the Creatures,* in which the praise and adoration of God cannot be conceived of as arising within us in any other way but in and through the living world. But Francis also knew what it meant to dwell deeply within the mystery of darkness and suffering. His is not the language of the dark night. But from his intimate encounter with lepers early on in his life, something that utterly transformed his sense of what it meant to be a follower of Christ, to the intense physical suffering he endured late in his life (malaria, trachoma, and a gastric ulcer so acute that it caused him to vomit blood), to the final harrowing experience of receiving the stigmata at La Verna, we see how far and deep into this mystery he was willing to travel.

A sense of great joy suffuses *Laudato Si',* as it does the life of St. Francis. But there is also an honest reckoning with suffering, darkness, and loss. We are being called to cultivate a capacity to behold and delight in all beings in the living world as our own brothers and sisters and to live accordingly. And we are being called to open ourselves to all that is broken—in us and in the world— and to consider what it will mean for us to turn seriously toward the work of healing. This deeply ethical vision of life is rooted in an understanding of the significance of what it means to see. Not seeing in a detached or possessive way, nor objectifying what one beholds; rather, cultivating a gaze that enables us to sense and become responsive to the intricate bonds that connect us to one another and to all living beings. A way of seeing that helps us recognize how deeply involved we are with the life and welfare of all sentient beings.

This entails what Pope Francis calls an "ecological conversion." The reference here to conversion is significant. Often associated with a conscious, rational choice to move from a non-religious position to an explicitly religious orientation, the word in fact connotes something deeper and more universal: a deep change of heart reflecting a transformed awareness of one's very identity in relation to everyone and everything. Such conversion, Pope Francis suggests, "entails the recognition of our errors, sins, faults and failures, and leads to a heartfelt repentance and desire to change."[6] And it "entails a loving awareness that we are not disconnected from the rest of creatures, but joined in a splendid universal communion."[7] This vision, while deeply personal in character, does not ignore the larger and deeper structural injustices that are such a critical part of our current environmental crisis, something to which *Laudato Si'* attends throughout. Still, this language suggests a wholehearted transformation of life, both personal and communal, rooted in a radical renewal of consciousness to which every human being is called in the present moment. The cultivation of a "loving awareness," then, is critical to the work of renewing our relationship with the created world, and rediscovering our own deep communion with the world.

It is also costly. Toward the beginning of the encyclical Pope Francis urges us to take seriously what it will mean to become "painfully aware" of the suffering we behold in the world and to allow it to become part of our own suffering.[8] "Our goal," he says, "is...to become painfully aware, to dare to turn what is happening to the world into our own personal suffering and thus to discover what each of us can do about it."[9] This, I would suggest, is the power of the contemplative gaze. It enables us to see deeply into the sacramental beauty of all that exists and to celebrate joyfully the gift that this beauty gives to us. It helps us cultivate an attitude of simple appreciation and gratitude for all that we have been given, and to respond wholeheartedly. And it makes possible a humble, empathetic engagement with the loss and suffering and darkness we encounter in our life in the world.

## "A Captivating Brightness Held and Opened": Learning to See

I began this chapter with a story about building a miniature house with my daughter on a remote beach on the Lost Coast of Northern California. I want to return to it for a moment to inquire about its potential meaning for our common work of learning to behold and cherish the living world.

Seeing is so deeply embodied: this is the realization that came to me that day laying on the black sand beach with my daughter at Bear Harbor. I am coming late to this realization, I know; but it has changed so much for me. For too long I have lived with a perception that the body—indeed material reality as a whole—was an impediment to authentic spiritual awareness. A residual after-effect of the narrowly Platonist or dualistic impulses that had seeped into the Christian spiritual tradition? Perhaps. But, however these ideas came to me, I fell victim to them more than I realized. And this made it difficult to accept and live within the fundamentally sacramental character of my own spiritual tradition or to experience and revere the world itself as sacramental. The fundamental truth of the incarnation—namely that all matter shimmers with transcendent beauty and power by virtue of Christ, the word of God having become "enfleshed" in a human body—largely eluded me. It should be noted that the Christian idea of the incarnation—God entering into mortal flesh through the person of Christ—has often been reduced to a theological idea with significance primarily for personal salvation. However, the early Christian community always understood this idea as having cosmic, or as we might say nowadays "ecological," significance. That is: God does not stand aloof from ordinary matter, but through the incarnate word is found in every element of creation. Through a long process of rehabilitation, I eventually "returned to my senses" and learned to stand more fully within my own body and reclaim this truth. And I slowly reclaimed a language, long known to the Christian spiritual tradition, for honouring this embodied form of awareness. The ancient Christian tradition had early on learned to speak of such awareness in terms of what it called the "spiritual senses," born of a recognition of how our five senses guide (metaphorically and actually) and shape our spiritual practice and understanding.[10] But the understanding of what it might mean to cultivate a spiritual practice attuned to our fully embodied life in the world was still emerging for me. That day on the beach with my daughter, I rediscovered something important: the pleasure of touch and the sense of involvement and joy—and yes, awareness—that it brings.

The coarse black sand between my fingers, slightly sticky from the salt water. The feeling of my hand moving slowly along the sand, smoothing a place where we will build our little house. Back and forth my hand moves, a humble but functional little tool. I pause from time to time to pick out bits of driftwood and kelp. Little pebbles. The simple sensual pleasure of this work. Also the pleasure of proximity. I have lowered my lanky frame down onto the ground; it is the only way I can possibly do this work. This ground's-eye view is a little strange: I am hardly ever down here (though as a child I used to crawl around in the dirt more or less constantly). I have forgotten how it feels to be this close to things, to see things from this intimate perspective. Occasionally, I pause in the work and roll over onto my back to gaze up at the sky; or lay on my side, face pressed hard against the sand and look out at the wild, churning ocean. These different ways of looking out at the world become part of the game. And the feeling of the earth upon my belly, the wind tousling my hair, my hands immersed in the beautiful texture of this dark, pulverized volcanic soil, and the presence of my daughter laughing beside me—all of these things converge into a single, embodied perception of reality. Suddenly, the habitual perception of myself as the principal actor in my own existence shifts. Yes, I am acting upon the world. But it is also acting on me. I am in the world. But it is also moving in me. There is no inside and outside. There is only this mysterious whole within which I live and move and have my being (Acts 17:28).

This awareness, I come to realize, is mediated through touch. The simple, visceral sensation of my own embodied self immersed in this place gives me a knowledge of myself as a living being among other living beings alive in the world that I can come to in no other way. It also makes possible a renewed awareness of the depth of loss that marks this place. I am, I realize, moving through an ecologically and morally compromised landscape: the stumps of ancient redwoods, long since cut down for timber, in their place vast stands of eucalyptus trees imported from Australia to provide wind breaks and firewood. Innumerable patches of scotch broom steadily encroaching on the native grasses. And, yes, the memory of those human beings who first inhabited this place, long ago killed or driven from their home: their blood is in the soil.

How to hold all of this in my awareness? How to live responsibly in relation to this beautiful, wondrous, wild place I have come to love but which is also marked by our collective failure to practise paying attention to the lives of other living beings? I am not sure. I think again of Pope Francis's call to become "painfully aware." It is an invitation to locate ourselves fully and deeply in the world and to take seriously our spiritual and ethical relationship with it. Here, too, I believe, we are being invited to see more deeply, more honestly, to overcome our habitual blindness.

I recall here the insight of the great Polish poet Czeslaw Milosz on what it means to open our eyes and look honestly at our own existence, our world, our shared history, everything:

"To see" means not only to have before one's eyes. It may mean also to pre-serve in memory. "To see and describe" may also mean to reconstruct in imagination. A distance achieved thanks to the mystery of time must not change events, landscapes, human figures into a tangle of shadows growing paler and paler. On the contrary, it can show them in full light, so that every event, every date becomes expressive and persists as an eternal reminder of human depravity and human greatness. Those who are alive receive a man-date from those who are silent forever. They can fulfill their duties only by trying to reconstruct precisely things as they were by wrestling the past from fictions and legends.[11]

Milosz's observations here are marked profoundly by the very particular losses he and so many of his countrymen endured during and after the Second World War in Europe. But we can, I think, consider their larger implications for us during this time of deepening environmental destruction. And we can extend this insight to ask how such a profoundly moral understanding of what it means to see can help us to cultivate an authentically contemplative response to the living world. How can we learn, again, to see?

Practice is critical. The embodied knowledge and awareness that comes to us through touch or smell or hearing or tasting or seeing is refined and deep-ened through repeated practice. This is of course common sense. But I wonder how seriously we take it? And I wonder also: To what embodied practices are we willing to give our attention and care? In particular, I wonder about those practices that are so simple and humble that they seem to fall below the thresh-old of our care and attention. I have already offered one example from my own experience. Let me offer you two more.

The Brazilian poet Carlos Drummond de Andrade has written a wonderful, whimsical poem—"No Meio do Caminho" (In the Middle of the Road)—that touches deeply on this question. "In the middle of the road there was a stone." This first line is repeated several times, in different modulations, with a strange rhythmic sameness. Only once is the rhythm broken, when the poet declares: "Never should I forget this event / in the life of my fatigued retinas."[12] It is difficult not to smile a little at hearing these words. Seriously? "Uma pedra!" A stone. A stone in the middle of the road. An encounter with a stone in the road as revelatory event. There is certainly whimsy here. But there is more than this. There are those fatigued retinas. The sense of a lifetime spent looking and looking and gradually forgetting how to look, growing weary of the whole prospect of looking at anything, of beholding anything with wonder, delight, or even mild interest. Becoming blind. And then suddenly there is a stone in the middle of the road. And against all expectation, wonder returns, delight is rekindled. There is a stone in the middle of the road. Unforgettable: when all seeing has come to an end, there is a small miracle—the prospect of beholding the world with new eyes.

I think also of the experience described in Seamus Heaney's poem "Ballynahinch Lake."[13] Stopping with a companion on a Sunday morning in Connemara. Gazing out at water birds splashing up on and down on the lake. That is all. But upon reflection, one can say (the poet says): something happened. Something moving and strange and potent. Here we encounter a rich, dense evocation of experience, of what it means (and feels like) to see the world. And how the gaze we cast upon the world returns to us, enters us, changes us. A simple moment: two people out for a drive in Connemara, pausing for a time "in the spring-cleaning light" by the side of a lake. That light: "a captivating brightness held and opened." And the "utter mountain mirrored in the lake" that "entered us like a wedge knocked sweetly home / Into core timber." And that pair of water birds, "air-heavers, far heavier than the air." The simple effort to describe what the eyes behold, but not only that: what it *feels* like to behold these things. The effort to conjure in language what it feels like to be held by your own vision of things. Not so simple, really. But what a work. "Something in us," the poet notes, referring to those "air-heavers," "had unhoused itself at the sight of them." Perhaps this is all that can be said: "something." How otherwise to express that fleeting but utterly gripping sensation of feeling yourself addressed, taken hold of, entered by another being? It is like being "unhoused." Like being caught and held in ecstasy, delight, wonder. Which helps to account for that slight pause just at the end when it is time to start the car and turn for home. A moment's hesitation to acknowledge the strange, wild beauty that has entered their lives. The recognition, expressed with wonderful understatement, that "this time, yes, it had indeed / Been useful to stop." A moment of simple awareness and awe: something that can build and deepen into a sustained contemplative practice.

## Seeing as a Tactic for Transformation and Resistance

These are simple and personal examples of what can happen when we slow down enough to notice the world around us; of what can happen when we begin to take in our own experience of the world. Small instances of *seeing*. Quite small. So much so that one might feel, especially in light of the immense and acute environmental crisis we are facing, that they are in fact too modest to be worthy of our attention. One feels the truth in this. Still, I want to suggest—paradoxically perhaps—that a commitment to learning how to see has never been more important. And that learning to cultivate the art of attention, or what ancient monks and philosophers called *prosoche*, will be critical to our and the world's future well-being.

This is nothing more or less than a claim for the enduring significance of contemplative practice to the work of deepening our ethical commitments, our social practices, our political life, and yes, our sense of ecological responsibility. Is this a quixotic claim? Perhaps. Still, learning to see, in a way that enables one to hold both the minute particulars and the whole in one synthetic gaze, has always been one of the primary practices of the great religious traditions. So, too, with ancient Christianity. Nor has this work of cultivating this contemplative gaze ever been understood (in its deepest sense) as something that can be divorced from our commitments to the larger community (whether this be understood in terms of the human community or the wider community of all living beings). The art of contemplative practice is at its heart holistic and inclusive. Which means that it has something distinctive and important to contribute to helping us recover a holistic ecological vision and practice.

There are two aspects of this contemplative work to which I want to draw attention here. The first is the particular contribution contemplative practice can make in helping us to cultivate a vision of the whole, or what Pierre Hadot (following ancient Greek philosophical thought) calls "cosmic consciousness." The second, more practical in character, is the contribution contemplative practice can make in helping us develop what Michel de Certeau calls "tactics" of resistance—ordinary practices that allow us to resist the hegemonic claims of the large, impersonal forces of the world and to live out of an alternative vision of life and reality that actually contributes to our and the world's well-being.

In a remarkable essay entitled "The Sage and the World," Pierre Hadot describes the kind of consciousness of the whole that was once common among sages, philosophers, and contemplatives but which we have to a great extent lost. At the heart of this consciousness was an experience of what he calls "unitive contemplation" rooted in an intense awareness of the present moment. "In order to *perceive* the world," Hadot notes, the sages believed that "we must, as it were, perceive our *unity with* the world, by means of an exercise of concentration on the present moment."[14] This exercise of concentration on the present moment is critical and is one of the most significant contributions of contemplative traditions of thought and practice to our current situation. For without a capacity to dwell deeply in the present instant, to perceive and become immersed in what we perceive, to enjoy and delight in it, we risk losing the precious reality we seek to behold. We risk losing both ourselves and this reality under the flood of preoccupations and projects for the future that come to dominate our consciousness. And we end up living not in the present moment but in a future that may never come. How elusive the present moment can seem. How oppressed we often feel by the tyranny, the relentless march of time. How much of our experience is eclipsed by, lost beneath, the breathless, merciless power of *chronos*. And we live, most of us anyway, under the cloud of this small, impoverished, and increasingly harried existence.[15] In light of this, the effort to reclaim the present moment can feel almost heroic. Even the smallest of moments, when noticed, held, and cherished for their own sake, become little miracles, celebrations of the simple joy of existing, the joy at beholding the world in all its mysterious beauty. But it is not easy to achieve. What is required to realize this depth of awareness, Hadot suggests, is nothing less than a "total transformation of our lives," a conversion of heart and mind so profound that we are no longer subject to the demons of worry and anxiety. Instead, we find ourselves able to behold ourselves and the world as part of a single unity.

What would it mean for us to practise this awareness in every moment of our lives? What would it mean to behold the world from such a perspective of deep awe, trust, and gratitude? How might it change the way we live and respond to the world we have been given? I realize that such questions may (again) feel too detached from the harsh reality of struggle and suffering we face every day in the contemporary world. I realize also that, while I am giving particular attention here to a practice that seeks to cultivate awareness of the simple beauty of our existence, our life in the world is far more than simple beauty. Still, if part of what we are struggling against in the present moment is an increasingly strong sense of alienation from the living world, a loss of sensitivity to its complex, intricate, wondrous reality, then it seems that part of our common work must be devoted to recovering the basic capacity to see and appreciate things—for their own sake, but also for all that they mean to us.

Pope Francis's recent encyclical places great emphasis on the need for such a transformation of awareness, and connects it intimately to the work of justice that our environmental crisis calls us to undertake. Ecological conversion, he notes, "entails a loving awareness that we are not disconnected from the rest of creatures, but joined in a splendid universal communion."[16] This, according to Francis, is what we are called to most deeply. In some ways, it occupies the very heart of *Laudato Si'*. But how to cultivate such awareness? How to deepen it? He suggests that we are being called in the present moment to rethink our whole approach to the idea of "quality of life," to open ourselves to a "prophetic and contemplative lifestyle, one capable of deep enjoyment free of the obsession with consumption." Instead, there is a new opportunity before us to learn: "To be serenely present to each reality, however small it may be, opens us to much greater horizons of understanding and personal fulfillment."[17] Nor is this simply an indulgence on our part, since "by learning to see and appreciate beauty, we learn to reject self-interested pragmatism."[18] Here spiritual and ethical practice converge: our capacity to see more fully and deeply the world we inhabit contributes to our capacity to love and care for it.

## Never Weary of Gazing

What does such seeing look like? Let me cite three examples that suggest what it feels like to notice ourselves as part of a "universal communion." The first is from a journal entry of Henry David Thoreau's, made on July 7, 1845, concerning something he calls "strange affinities":

> There are strange affinities in this universe—strange ties stranger harmonies and relationships, what kin am I to some wildest point among the mountains—high up ones shaggy side—in the gray morning twilight draped with mist—suspended in low wreathes from the dead willows and bare firs that stand here and there in the water, as if here were the evidence of those old contexts between the land and water which we read of. But why should I

find anything to welcome me in such a nook as this—This faint reflection this dim watery eye—where in some angle of the hills the woods meet the water's edge and a grey tarn lies sleeping.[19]

There is a haunting familiarity and pathos to the questions at the heart of this observation: "What kin am I" to all these beings? Do we still feel these "strange affinities" that bind us to the lives of others? What is it to become immersed in the mood of a place, which suffuses us, becomes part of our own mood, circling around and moving through us until we no longer know where the boundary exists between ourselves and another? "Is there anything to welcome" us here?

We have learned to be dismissive of such questions: they are too sentimental or romantic or anthropomorphic. And yet the feeling persists that these "strange affinities" are real and that our own lives are bound, deeply and irrevocably, to the lives of other beings. There is a deep moral meaning to these affinities. So asserts American poet and essayist Mary Oliver:

I would say that there exist a thousand unbreakable links between each of us and everything else, and that our dignity and our chances are one. The farthest star and the mud at our feet are a family; and there is no decency or sense in honoring one thing, or a few things, and then closing the list. The pine tree, the leopard, the Platte River, and ourselves—we are at risk together, or we are on our way to a sustainable world together. We are each other's destiny.[20]

There is an echo here of Pierre Hadot's observation regarding the sensibility of the ancient sages: "In order to *perceive* the world we must...perceive our *unity with* the world." *Prosoche*: attention, awareness, openness to everything.

What kind of work or practice is this? And where does it begin? Returning to *Laudato Si'*, we catch a glimpse of what it might mean: "Each reality, however small," is worthy of our attention and care. So: begin with the palpable particular. Practise paying attention, being open, aware. Notice what is unfolding before you. Cherish it.

One catches a vivid glimpse of this sensibility in John Muir's account of the trees he encountered during his first summer in the Sierra Nevada mountains. He notes carefully the distinctive qualities of particular specimens of goldcup oak, Douglas spruce, yellow pine, silver fir, and sequoia. But it is the sugar pine that seems to have caught and held his attention most deeply. He called it "an inexhaustible study and source of pleasure":

> I never weary of gazing at its grand tasseled cones, its perfectly round bole one hundred feet or more without a limb, the fine purplish color of its bark, and its magnificent outsweeping, down-curving feathery arms forming a crown always bold and striking and exhilarating.... At the age of fifty to one hundred years it begins to acquire individuality, so that no two are alike in their prime or old age. Every tree calls for special admiration. I have been making many sketches, and regret that I cannot draw every needle.[21]

It is not easy to know what to make of this expression of simple, stripped-down attention to the particularity of an individual being. The work of perception here feels almost unimaginatively slow, even tedious. We no longer look at things this way, certainly not this carefully, or for this long. "I never weary of gazing," Muir says, sitting before the old sugar pine. And there he sits making sketch after sketch, full of regret that he cannot draw every needle.

"I never weary of gazing." Such a sensibility was rare then and it is rare now. I think we can learn much from this practice of simple attention, not least how to free ourselves from the obsessions of consumerism and to cultivate instead what Laudato Si' calls "the capacity to be happy with little." There is a reversal here, an inversion of the widespread Western consumerist emphasis on the value of accumulation. "It is a return to that simplicity which allows us to stop and appreciate the small things, to be grateful for the opportunities that life affords us, to be spiritually detached from what we possess, and not to succumb to sadness for what we lack."[22] Pope Francis's own gaze is here fixed primarily on those living in the Global North, where materialism has become a moral-spiritual problem. "It is not a lesser life or one lived with less intensity," he notes, perhaps anticipating the kinds of objections that might be levelled against such a view:

> On the contrary, it is a way of living life to the full. In reality, those who enjoy more and live better each moment are those who have given up dipping here and there, always on the look-out for what they do not have. They experience what it means to appreciate each person and each thing, learning familiarity with the simplest things and how to enjoy them.[23]

This moment-by-moment practice of attention can, I think, be helpfully understood in terms of what Michel de Certeau calls a "tactic." In his book *The Practice of Everyday Life*, Certeau makes a distinction between "strategies" and "tactics." Strategies, he says, are designs on the world employed and enacted by those with power—corporations, governments, and other entities with sufficient power to enact a particular strategy in the world. Tactics, on the other hand, are employed by those with little or no power, and often represent an effort to resist the tyranny of various strategies imposed upon them from the outside, without their assent. Certeau notes that it is often in the most ordinary practices of everyday life— cooking, walking, gardening, making art—that such tactics come to expression. Those employing them are exercising their own agency in the world and giving expression to their own vision of life. In this sense, these often modest tactics come to have a subversive character.[24] They reflect and embody a resistance to the dominant forces under whose strategic power we often live our lives. They point to the possibility of living with freedom and joy amidst circumstances that might otherwise seem utterly bereft of them. And they can create the conditions for movements of political resistance that challenge the dominant strategies that threaten to reduce our lives and the life of the world itself to a wasteland.

Can the art of contemplative practice be considered such a tactic? Can it help us to recover a sense of the world as a precious gift to be cherished and cared for and perhaps contribute to the practical work of political, cultural, and ecological transformation that we so desperately need in the present moment? I believe it can and should.

## Rebuilding Our Common Home

I return now at the end to that little house near the ocean. The memory of sharing that sweet moment with my daughter remains precious to me. It was a moment in which we neither needed nor wanted anything beyond the simple enjoyment of the work we had been given to do, the place in which it unfolded—the black sand beach, the crashing waves, the crying gulls, and the afternoon sunlight pouring down—and of course one another. It was, and is still, a little emblem of paradise. And that miniature house, over which we exercised so much care and imagination. As if it mattered—which of course it did, and does. More than we knew, I suspect, we wanted to make something beautiful, a little dwelling where creatures like us could live free from worry and anxiety. Where we could live in that little paradise unending.

This desire to make things, to create a world and to inhabit it—where does it come from? I do not know. I only know that there is a pleasure, deep and pure, that comes from making something beautiful, fitting stones into a pattern, laying a floor, creating a garden, making a life. Even if the making is all there is, even if the thing made is ephemeral and not destined to last, there is pleasure and joy in the making. Sometimes, though, our creations *do* endure. We are able to behold the work and feel it exert its magic on us, kindling the imagination, taking us out of ourselves, if only for a moment, into another world.

That is how I think of it now, anyway. I realize, too, how lying flat on the ground, working with those small stones and pieces of driftwood and kelp, fitting the pieces carefully into place, conversing lazily with my daughter about the little world we were creating, we forgot about time. We became lost in the spaciousness of that moment. There *was* only that moment. Nothing else. And then there was something beautiful, a house, with doors, windows, a roof, encircled by a low fence and a few trees, a glistening path leading to the entrance. So inviting! Our little home. I can see it so clearly, gleaming in the afternoon light.

I can also see, if I look hard enough, the place Sally Bell and her people called home: faint, hidden in the shadows, and marked by blood. But still alive and real, especially among her descendants and among all who have come to love and cherish this place.

And I behold St. Francis and recall his own rebuilding project at San Damiano. How he responded so wholeheartedly to the call to rebuild a church that had fallen into ruins. How with wonderful naïveté and directness, he began reassembling the stones of the San Damiano church itself: his own little house. And so began the larger work of reparation, including his joyful rekindling of relationship with all living beings. His *Canticle to the Creatures*, composed toward the end of his life, by which time he had spent himself completely on this work of restoration, reminds us of the deep intimacy with the world that is possible for us. With disarming simplicity, he invites us to pause and see and feel the truth that every living being is kin to us. And that we can only love and honour God fully *in and through* our love and care for the elements of creation—praise to you, oh God, *through* brother sun, *through* sister moon, *through* brother wind and air, *through* sister water. Such intimacy and joy. A great vision and a great challenge.

This is our only home. In this moment of acute environmental crisis, we are once again being called upon to see all that we have been given, to care for our common home. To help repair it and restore it to health. Can we learn again to see all that we have been given? To cherish, delight in, and care for it? We can and we must.

## Notes

1 The Sinkyone Wilderness is a California State Park located on the southern portion of the Lost Coast a few miles north of Mendocino.

2 Benjamin Madley, *An American Genocide: The United States and the California Indian Catastrophe, 1846–1873* (New Haven, CT: Yale University Press, 2016), 210–11.

3 For an early and influential account of the relationship between ecology and virtue ethics, see Louke van Wensveen, *Dirty Virtues: The Emergence of Ecological Virtue Ethics* (Amherst, NY: Humanity Books, 1999). On the relationship between environmental ethics and spirituality, see Willis Jenkins, *Ecologies of Grace: Environmental Ethics and Christian Theology* (New York: Oxford University Press, 2008), esp. 93–111.

4 On the idea of "contemplative ecology," see Douglas E. Christie, *The Blue Sapphire of the Mind: Notes for a Contemplative Ecology* (New York: Oxford University Press, 2013).

5 *Laudato Si': On Care for Our Common Home* (Washington, DC: United States Conference of Catholic Bishops, 2015), sec. 139.

6 *Laudato Si'*, sec. 218.

7 *Laudato Si'*, sec. 220.

8 *Laudato Si'*, sec. 19.

9 *Laudato Si'*, sec. 19.

10 Paul Gavrilyuk and Sara Coakely, eds., *The Spiritual Senses: Perceiving God in Western Christianity* (Cambridge: Cambridge University Press, 2014).

11 Czeslaw Milosz, *Beginning with My Streets: Essays and Recollections* (New York: Farrar, Strauss and Giroux, 1991), 280–1.

12 Carlos Drummond de Andrade, "No Meio do Caminho" [In the Middle of the Road], *Complete Poems: 1927–1929*, trans. Elizabeth Bishop (New York: Noonday Press, 1983).

13 Seamus Heaney, *Electric Light* (New York: Farrar, Straus and Giroux, 2001).

14 Pierre Hadot, *Philosophy as a Way of Life*, ed. Arnold I. Davidson (Oxford: Blackwell, 1995), 261.

15 David Steindl Rast, "Thomas Merton, Man of Prayer," in Thomas Merton, *Monk: A Monastic Tribute*, ed. Patrick Hart, OCSO (Kalamazoo, MI: Cistercian Publications, 1974), 79–89.

16  *Laudato Si'*, sec. 220.

17  *Laudato Si'*, sec. 222.

18  *Laudato Si'*, sec. 215.

19  Henry David Thoreau, *Journal*, Volume 2: 1842–1848, ed. Robert Sattelmeyer (Princeton: Princeton University Press, 1984), 158.

20  Mary Oliver, *UpStream* (New York: Penguin, 2016), 154.

21  John Muir, *My First Summer in the Sierra* (New York: Penguin, 1987), 51.

22  *Laudato Si'*, sec. 222.

23  *Laudato Si'*, sec. 223.

24  Michel de Certeau, *The Practice of Everyday Life*, 3rd ed. (Berkeley: University of California Press, 2011).

# Part IV

# PHILOSOPHIES *of* VIRTUE ETHICS

# EIGHT

# The Ethic of Sustainable Well-Being and Well-Becoming

## A Systems Approach to Virtue Ethics

THOMAS FALKENBERG

## Preamble

ECOLOGICAL FOOTPRINT RESEARCH MEASURES "THE DEMAND [human] populations and activities place on the biosphere in a given year" relative to the "biologically productive land and sea area available to provide the ecosystem services that humanity consumes."[1] In the most recent *Living Planet Report* we read that "humanity currently needs the regenerative capacity of 1.6 Earths to provide the goods and services we use each year."[2] In other words, humans are consuming 60 percent over what the global ecosystem is able to sustainably provide. Ecological footprint research and other ecosystem services research[3] document the ever-increasing rate of destruction of the ecological support systems needed to sustain human and non-human life. At the same time, we have an inequitable distribution of means across, but also within, nations to address subsistence and a number of other fundamental human needs: for example, the distribution of financial resources,[4] the distribution of food supplies,[5] and the distribution of ecosystem services.[6] The influential definition given by the World Commission on Environment and Development describes sustainable development as "development that meets the needs of the present without compromising the ability of future generations to meet their own needs." This definition accurately depicts what humanity is currently not doing.

This unsustainable development and the resulting ecological crisis have been problematized from different perspectives. For instance, deep ecologists[7] address the ecological crisis centrally as an *ontological* problem—namely, as a crisis of the way we structure our earthly household (ecology) as a hierarchy of beings, with humans at the top (anthropocentric ontology) giving their perceived needs priority over everything else. Ecological economists,[8] meanwhile, generally address the ecological crisis as a crisis of *pragmatic* nature, concerning the way we run our household (economy), the way we treat the means for human living. In this chapter I will address the crisis as an *ethical* problem by proposing an ethic of sustainable well-being and well-becoming. This approach is aligned with the one adopted by ecological virtue ethicists,[9] although, as we will see, the ethics of sustainable well-being and well-becoming take virtues not as foundational but rather as derived. Ethics as a field of inquiry is often seen—and I adopt this view here—as being concerned with "questions about human flourishing, about what it means for a life to be well lived."[10] An ethic of sustainable well-being and well-becoming, then, is a guide to living a flourishing life that has one guiding principle: to live *sustainably* well. Virtues are understood in virtue ethics as dispositions that help us live a flourishing life. The virtues relevant to an ethic of sustainable well-being and well-becoming are derived from the concepts of sustainable well-being and well-becoming. An ethic of sustainable well-being and well-becoming identifies what dispositions count as virtues that make the ethics of sustainable well-being and well-becoming a virtue ethics—not the other way around.

In developing the ethics of sustainable well-being and well-becoming in this chapter I proceed as follows. First, I draw on Falkenberg[11] to outline a framework for conceptualizing individual human well-being and well-becoming, which takes its starting point in the notion that human beings are first and foremost biological systems. This framework serves as the theoretical and conceptual basis for the integrated systems approach to the ethics of sustainable well-being and well-becoming. Second, I systemically interconnect humans as individual biological systems to each other through social systems and to other living organisms through ecological systems. Both systems are integrated into socio-ecological systems and, drawing on approaches in the social-ecological research field, I explicate the idea of a sustainable development of socio-ecological systems. Third, I link the concern for individual human well-being and well-becoming developed in the first section and the concern for a sustainable development of socio-ecological systems developed in the second by developing the concept of *sustainable* well-being and well-becoming (human flourishing). This concept provides, then, an answer to the ethical question of how I should live my life as a human being, which leads, fourthly, to the development of an ethic of sustainable well-being and well-becoming. Finally, I explicate the role that virtues play in an ethic of sustainable well-being and well-becoming.

## Individual Well-Being and Well-Becoming: An Integrated Framework Approach[12]

### Humans as Biological Systems

Drawing on Maturana and Varela[13] and those who have followed their path of a theoretical view of life as an autopoietic biological system,[14] I take the view that a human being is a multi-cell autopoietic system with a human-typical organization of the system, and that the human organization consists of system components and their relationships, like the heart pumping blood through arteries to carry oxygen to muscle cells. The different systems that make up a human being (like cells or larger systems like the nervous system) are structurally coupled, meaning that "there is a history of recurrent interactions [between the different structures] leading to the structural congruence between two (or more) systems."[15] Intentionally lifting my arm is possible because of structural coupling between my cognitive system and my sensory-motor system. Compared to the organization of a human "system," the *structure* of a human system is temporally and spatially specific to each human, and changes over time. These changes, whether at the cell level or at the whole human being level, are triggered, but not determined, by perturbations of the human

organism occasioned by the living system's environment. While such perturbations trigger a change, which specific change occurs within the autopoietic
system is determined by the structure of the living being at the time of the
perturbation. Simultaneously, structural changes can be initiated from within
the system itself. For example, "remembering" something as an event within
a human's nervous system might lead to changes within the system that make
a "remembering event" of this type more common. Often, structural changes
in a human organism are what we would call "learning." The history of a single
human being consists of ongoing structural changes.

The view of humans as living organisms impacts the framing of a conceptualization of human well-being and well-becoming (henceforth, human flourishing) in three ways. First, it suggests that human flourishing is ultimately to be
understood at the level of the individual being, because the structure of each
human being, which is the basis for how a human responds to inner and outer
perturbations ("experiences"), is distinctly individual. If the system structure
and, thus, the structurally determined responses to system-external perturbations are specific to each individual human being, then a conceptualization of
human flourishing that has to make use of these two aspects of being human
(structure and responses) is best done at the level of the individual human being.
Second, the view of human beings as biological living systems also suggests a
radical constructivist epistemology argued for by Ernst von Glaserfeld, which
is built upon the following two principles: "knowledge is not passively received
but built up by the cognizing subject," and "the function of cognition is adaptive
and serves the organization of the experiential world, not the discovery of ontological reality."[16] Thus, a conceptualization of human flourishing needs to give
consideration to the cognizing beings' subjective experiences. Third, the view of
humans as autopoietic systems suggests a *systemic* approach to flourishing. This
implies in particular that human flourishing as a quality of a system has to be
understood as a *relational* quality—namely a quality of the system that relates a
human being (as a system) to its environment, to which it is structurally coupled.

The scholarly literature on human flourishing is vast, quite diverse, and
spread across different academic disciplines.[17] While the approaches are by no
means equivalent, they point to different traditions in addressing the common
foundational idea: that humans care deeply about and for their own and other
people's flourishing. The approach I propose in this chapter is aligned with this
common foundational idea, which I have expressed elsewhere as follows:

In other words, *the concept of well-being is to capture what humans aim for when they exert their agency to live their lives one way rather than another.* This concept of well-being has the quality of "prospectivity"... or future directedness.... This identifies one central reason for the importance of the concept of well-being: What we conceptualize it to mean can and should direct our decisions and actions at the individual, socio-cultural, and socio-political level.[18]

According to Martin, Sugarman, and Thompson, human agency emerges from, constantly interacts with, and is continuously shaped by the socio-cultural environment into which such agency is embedded.[19] Hence, "individual" flourishing has to be understood in such situated emergence from its socio-cultural environment.

To make the conceptualization of individual flourishing practically useful, this foundational idea needs to be made more specific. Since humans are first and foremost biological organisms, such a specification seems to be best undertaken by identifying qualities of what it means to be well for humans as living organisms. The concept of individual flourishing I develop in this section identifies the following five qualities of being alive as a human being as defining components of the concept of individual flourishing:

- having agentic capabilities linked to fundamental human needs;
- experiencing situational opportunities to engage one's agentic capabilities in relevant life domains;
- enjoying life;
- living a meaningful life; and
- experiencing personal and communal connections that contribute to one's well-being and well-becoming.

These components are to capture what drives us humans when living our lives one way rather than another: that is, we want to have agency and the required capabilities in relevant life domains to address our fundamental human needs; we want to enjoy life; and so on.

The five components of human well-being and well-becoming identified above draw on different traditions of human flourishing, making my proposal an *integrated* approach in the sense that it integrates these different traditions. Here I follow Gordon Allport, who, with reference to William James, suggests:

"By their own theories of human nature psychologists have the power of ele-
vating or degrading this same nature. Debasing assumptions debase human
beings; generous assumptions exalt them."[20] Bringing together different visons
of human flourishing in the presented approach augments our depiction of an
exalted human nature.

A third part of the core idea behind my approach to human flourishing is
that what is proposed is not the conceptualization of well-being and well-becoming itself, but rather a framework for such conceptualization. Such a framework approach accounts for the justified critique that has been levied against
the initial "trans-cultural" approach to the conceptualization of human flourishing in positive psychology (e.g., Seligman and Csikszentmihalyi[21])—for
instance, by Christopher, who warns:

> Understandings of psychological well-being necessarily rely upon moral
> visions that are culturally embedded and frequently culture specific. If we
> forget this point and believe that we are discovering universal and ahis-
> torical psychological truths rather than reinterpreting and extending our
> society's or community's moral visions, then we run the high risk of casting
> non-Western people, ethnic minorities, and women as inherently less psy-
> chologically healthy.[22]

Accordingly, the proposed framework approach provides only five general
components of individual well-being and well-becoming, while for any practical purposes—for instance for the purpose of assessing students' well-being
and well-becoming in a particular school division—the meaning of each of the
five components needs to be more specified in accordance with and to account
for a local and purposive understanding of human flourishing.

The next section builds on the presented integrated framework for conceptualizing individual flourishing toward a notion of *sustainable* well-being and
well-becoming.[23]

## Sustainable Well-Being and Well-Becoming

In this section I take the next step toward an ethic of sustainable flourishing
(well-being and well-becoming). I first conceptualize social-ecological systems

and what it means for such systems to be sustainable or to develop sustainably. I then link the notion of a sustainable social-ecological system to the concept of human well-being and well-becoming developed in the previous section to arrive at a conceptualization of sustainable flourishing—that is, what it means to live sustainably well.

## Social-Ecological Systems

*Social systems* are made up of human agents as their components and the patterns of their interactions (social practices) as their organization.[24] Humans' social interactions and their patterns give rise to, maintain, and transform the social structures of social systems, while on the other hand a given social structure conditions humans' social interactions. From a social systems perspective, human agency as part of the drive for individual well-being manifests itself in human beings in drawing (consciously or subconsciously) on rules within the social structure to engage in particular social practices to achieve intended purposes. This is where human agency and the underlying psychic system—and thus the pursuit of individual well-being—interact with the social system. For example, let us assume I wish to go for an enjoyable bike ride on a nice summer evening. In riding my bike, I follow the social practice of abiding by traffic regulations—for example, by riding on the right side of the road. My enactment of this social rule helps to reinforce others' following the same rules.

*Ecological systems* are systems that have living organisms as their components and the patterns of their interactions as their organization. As such, these components are higher-order unities in Maturana and Varela's sense,[25] and the notion of the ecological system is to capture the interaction of these higher-order unities. *Social-ecological systems* are systems that integrate the components of social and ecological systems and that have patterns of their interactions as their organization. To provide an illustrative example of the idea of a social-ecological system, I draw on Davis's contrasting of the human relationship with nature as understood in "ancient wisdom" communities and in "the modern world," respectively:

> Revered by Hindu, Buddhist, and Jain, [Mount] Kailash [in Tibet] is considered so sacred that no one is allowed to walk upon its slopes, let alone climb to the summit.... In Canada, we treat the land quite differently. Against the

wishes of all First Nations, the government of British Columbia has opened the Sacred Headwaters to industrial development.[26]

Environmental concerns aside, think for a moment of what these proposals [of opening the Sacred Headwaters to industrial development] imply about our culture. We accept it as normal that people who have never been on the land, who have no history or connection to the country, may legally secure the right to come in and by the very nature of their enterprise leave in their wake a cultural and physical landscape utterly transformed and desecrated.[27]

Davis contrasts here two distinct social-ecological systems with different patterns of interactions between the social and the ecological system involved. In the first case, people not connected to the area coming into the area and cutting down and removing the trees enact a social rule of that social structure, namely that such action is acceptable, even desirable and unquestionable. When people begin to see the "cultural and physical landscape utterly transformed and desecrated,"[28] the normalcy of the original social rule that strangers can go into an area and cut down the trees may be increasingly challenged, leading to different social rules guiding the interaction between humans and the components of the ecological system—in this case, trees. The reverence for Mount Kailash exhibited by Hindus and other religious groups identifies such a different social rule—namely, not walking on the slopes of a sacred mountain.

## Sustainable Development in Social-Ecological Systems

As living systems, humans are constantly perturbed by and interact with their respective environments, leading to constant changes. The same holds true for other living systems that make up the components of the earth-wide ecological system. For this reason, the concept of sustainable *development* is of greater relevance here than the concept of a sustainable system. The former is, of course, derived from the latter, but focusing on sustainable development gives consideration to the constant *process of becoming* that humans as living systems are engaged in from the beginning of their individual existence till death. Holling, one of the leading scholars in the social-ecological research field,[29] defines sustainable development for systems as follows:

Sustainability is the capacity to create, test, and maintain adaptive capability. Development is the process of creating, testing, and maintaining opportunity. The phrase that combines the two, "sustainable development," therefore refers to the goal of fostering adaptive capabilities while simultaneously creating opportunities. It is therefore not an oxymoron but a term that describes a logical partnership.[30]

In other words, sustainable development of the ecological system is development toward an expansion of the system's adaptive capability vis-à-vis constant inner and outer perturbations. Thus, neither "sustainability" nor "sustainable development" refers to a fixed end state because ecological systems are constantly changing in response to the perturbations they undergo, resulting in diverse types of living systems. The reduction of the diversity of types of living systems within our current earth-wide ecosystem is considered a decrease in the ecosystem's adaptive capability to respond to system-disturbing perturbations.[31] The sustainable development of a social-ecological system is thus conceptualized as the development of the respective social and ecological sub-systems toward an expansion of their adaptive capabilities to face constant inner and outer perturbations.

### Linking Individual Well-Being and Well-Becoming with Sustainable Development: Sustainable Well-Being and Well-Becoming

In this chapter, individual well-being has been conceptualized as a property of individual human beings as autopoietic systems. Sustainable development has been conceptualized as a property of systems with individual human beings as their components. In the case of ecological systems, these components are living beings, and in the case of social systems they are social agents. How can two very distinct types of properties be linked to a foundation of an ethic of sustainable well-being? The approach to linking the two system properties proposed here is as follows. The framework for the concept of individual well-being proposed above identifies aspects of being human—namely, the potential for living and lived experience—through which the quality of a person's life can be assessed. The concept of sustainable development of a social-ecological system outlined above identifies a property of a social-ecological system—namely, a development toward, and expansion of, the respective system's adaptive capability. Every human being is embedded into the earth-wide social-ecological system. The integrating concept of sustainable well-being qualifies individual well-being in such a way that its components are "supportive of" the sustainable development of the earth-wide social-ecological system. In other words, the components of the framework for individual well-being, like having agentic capabilities and enjoying life, are qualified in such a way that their concretization in an individually and culturally specific concept of individual well-being supports the development of greater adaptive capabilities of the earth-wide social-ecological system. I will call such integrated concept "sustainable well-being."

Developing the idea of what it means for the components of individual well-being to be "supportive of" the sustainable development of social-ecological systems is beyond the scope of this chapter, but I would like briefly to illustrate the idea nonetheless. Let's assume that diversity of life on earth is contributing to the adaptive capability of the global ecological system. If a person then lives a sustainably flourishing life, it implies that what makes the person's life enjoyable is supportive of diversity of life on earth. If "supportive of" means, at a minimum, "not in conflict with," then what makes my life enjoyable when living a sustainably flourishing life cannot involve the eating of those kinds of fish that are endangered, because a stark reduction of the diversity of ocean life means a reduction of the adaptive capabilities of the earth-wide ecological system.

Such an integrated concept of sustainable well-being would be a special case of individual well-being because qualifying the components of individual well-being in such a way that they support the development of the adaptive capability of the ecological system would add another criterion to each of the components. Conceptually, if someone is sustainably well, then that person would also be well in the sense of individual well-being. However, the two concepts—individual well-being and sustainable well-being—are quite distinct in terms of their rationalization for their importance in human living. The framework for conceptualizing the individual well-being of humans as living organism of a certain type was developed by drawing on different approaches to human well-being, each of which with a different understanding of the functioning of humans as specific living organisms. This framework finds its rationale in the assumption that individual well-being is what humans "naturally" aim for when exerting their agency. On the other hand, qualifying individual well-being so that each component is "supportive of" the sustainable development of the surrounding social-ecological systems goes beyond a view of "human nature," and adds a *normative* aspect to human well-being—namely, the concern for such sustainable development. The approach to human well-being proposed in this chapter will account for this categorical difference between individual and sustainable well-being by proposing the former as the way in which the approach conceptualizes human well-being and by proposing the latter as the core concept for an ethic of sustainable well-being, which will be developed in the next section.

## An Ethic of Sustainable Well-Being and Well-Becoming

With the concept of sustainable well-being explicated, I now turn to developing an ethic of sustainable well-being. I follow Appiah's distinction between ethics and morality: "I'll generally follow Aristotle in using 'ethics' to refer to questions about human flourishing, about what it means for a life to be well lived. I'll use 'morality' to designate something narrower, the constraints that govern how we should and should not treat other people."[32] Accordingly, an ethic of sustainable well-being would suggest that what counts for the ethical orientation of living one's life is sustainable well-being. In other words, in an ethic of sustainable well-being, we are to orient our agency toward living sustainably well.

How can we respond to the moral question of what we owe other living beings, including other humans, from the stance of an ethic of sustainable well-being? An ethic of sustainable well-being *frames* a response to the moral question in two ways, although it does not provide a specific answer to the question. First, living sustainably well means that I am concerned for the development of the adaptive capabilities of the social-ecological system(s) and, thus, for the quality of life of other living beings. For instance, a social-ecological system in which the substantial needs of all human beings are not adequately met makes the system far more vulnerable, thus challenging its adaptability. The questions—which aspects of the quality of life of others, and how are those affected by my concern for the adaptive capabilities of the social-ecological system(s)?—need to be addressed *outside* of an ethic of sustainable well-being. This means that my concern for the well-being of other beings is in this general sense implied by an ethic of sustainable well-being.

Second, in the way in which "individual well-being" is conceptualized, the quality of my relationships with other beings, including other humans, is crucial. For instance, "experiencing agentic situational opportunities to engage one's agentic capabilities in relevant life domains" and "experiencing personal and communal connections that contribute to one's well-being and well-becoming" are two components of the framework for individual well-being that are dependent on the quality of life of others. For instance, I might take the Golden Rule—probably the most prominent culture-transcending moral principle[33]—as my guiding moral principle in order to relate to my fellow humans in such a way that I indeed have opportunities to engage my agentic capabilities. Employing the Golden Rule as my moral principle might also support my experiencing of communal connections that contribute to my well-being.

At the beginning of this chapter I mentioned the role that dispositions play in the functioning of humans as autopoietic bio-psychic systems. In the next section I identify these dispositions as virtues and outline the role that they play in an ethic of sustainable well-being.

## Virtues in an Ethic of Sustainable Well-Being and Well-Becoming

### Virtues and "the Human Condition"

"Virtue ethics" is a label describing a cluster of ethical theories. Virtue-ethics approaches take as foundational character traits, or dispositions, that are identified as virtues. An ethically good person, then, is a virtuous person, and what is right to do in a given situation is what a virtuous person would do in that situation.[34]

In his discussion of different varieties of virtue ethics, Oakley specifies "two main kinds of approach taken by virtue ethicists in grounding the character of the good agent." The first is the "eudaimonistic view" of virtues, which he identifies as "the more prominent of these approaches."[35] This approach "draws on the Aristotelian view that the content of virtuous character is determined by what we need, or what we are, qua *human* beings."[36] According to Oakley, this first kind of approach comes in two versions, the first of which is proposing that virtuous living is necessary for eudaimonia, and the second version proposing that a virtuous life must be overall beneficial to the person living it.[37]

In the first version, "the virtues are character-traits which we need to live humanly flourishing lives":

> character-traits such as benevolence, honesty, and justice are virtues because they feature importantly among an interlocking web of intrinsic good… without which we cannot have Eudaimonia, or a flourishing life for a human being. Moreover, these traits and activities, when coordinated by the governing virtue of phronesis (or practical wisdom), are regarded as together partly *constitutive* of Eudaimonia—that is, the virtues are intrinsically good components of a good human life.[38]

According to this view, "acting virtuously is a constituent part of what a good human life consists in."[39]

The second version of this first kind of approach to virtue ethics grounds "the virtues not so much in the idea of a good human being, but rather in what is good *for* human beings."[40] Oakley, who acknowledges Philippa Foot[41] as "the best-known exponent" of this second version, explains that Foot "derive[s] virtues from what is beneficial to humans either individually or as a community."[42] Compared to the first version, in this second version living a virtuous life or being a virtuous person is not a constitutive part of a flourishing life but rather *contributes to* or *is a means for* living a flourishing life. Hursthouse, whose approach to virtue ethics is also an example of this second version, describes the relationship between being virtuous and living a flourishing life as follows: "The claim is not that possession of the virtues guarantees that one will flourish. The claim is that they are the only reliable bet—even though, it is agreed, I might be unlucky and, precisely because of my virtue, wind up dying early or with my life marred or ruined."[43] It is this second version of the first kind of approach to virtue ethics that is of interest here.

## *The Role of Virtues in an Ethic of Sustainable Well-Being and Well-Becoming*

In the approach to virtue ethics adopted here, virtues are defined by their role in and contribution to living a flourishing life. While virtues do not define what it means to live a flourishing life, they, as Hursthouse puts it, "are the only reliable bet" for living a flourishing life. This is exactly the role of virtues in the ethic of sustainable well-being outlined in this chapter: virtues are not defining what it means to live sustainably well; rather, virtues or character traits (dispositions) are understood here as "the only reliable bet" toward accomplishing living sustainably well. Hence, whether virtues play this role or not for living sustainably well is not an "empirical" question but rather a defining criterion. To what does this statement commit an ethic of sustainable well-being? First, such an ethic is a virtue ethics, at least in the specific sense of a virtue ethics as characterized above. Second, it also has implications for education for sustainable well-being. Such education needs to focus, for instance, not just on culturally specified capabilities relevant to participation in the cultural systems an individual is embedded in, but rather on such capabilities "in conjunction with" relevant virtues.

An ethic of sustainable well-being turns many educational approaches to virtue, as well as character education, on their heads. In these approaches, the virtue of temperance or moderation in drinking alcohol—to choose one example—is treated as a "guide" to follow, such that living by this guide is defined as living well. In the approach proposed in this chapter, on the other hand, what temperance means in the first place, and how it applies to the consumption of alcohol, depends on its contribution to a person's well-being.

How might virtues function as means toward becoming or being sustainably well? Following, again, Aristotle's general approach to virtue ethics, virtues are dispositions[44] to act and feel in particular ways in spheres of human experiences. As Nussbaum writes about Aristotle's approach to virtues:

> In the enumeration of the virtues in the *Nicomachean Ethics*, he [Aristotle] carries the line of thought further, suggesting that the reference of the virtue terms is fixed by spheres of choice, frequently connected with our finitude and limitation, that we encounter in virtue of shared conditions of human existence. The question about virtue usually arises in areas in which human choice is both non-optional and somewhat problematic. (Thus, he stresses, there is no virtue involving the regulation of listening to attractive sounds, or seeing pleasing sights.) Each family of virtue and vice or deficiency words attaches to some such sphere.[45]

In other words, virtues are those habituated principles that guide our actions and the emotions that arise in particular situations within a sphere of human

experience that is linked with the respective virtue. This understanding of virtues puts constraints on the role of virtues when assessing people's sustainable well-being because having virtues is not an indicator of living sustainably
well, but rather is a *means toward* the latter. Since the enactment of means is
no guarantee for living sustainably well, it would be inappropriate to use the
observation of virtuous behaviour as an indicator of how sustainably well the
life of a particular person is proceeding. In an ethic of sustainable well-being, the criteria have to come from the culturally specified components of
individual well-being, relativized to a sustainable development of the social-
ecological system.

Let me illustrate this role of virtues in an ethic of sustainable well-being
by drawing on temperance or moderation as an example of a classical virtue
in Western cultures. For Aristotle, temperance is the virtue that guides us in
our actions and feelings in the human sphere of experiencing "pleasures and
pain."[46] For instance, our feeling of excitement when we receive a wrapped
present exists in this sphere; similarly, our acting upon our urge to eat something salty by getting up from the couch, where we just saw a commercial for
potato chips on the TV, and going to the cupboard in the kitchen, where we
intend to get a still half-full bag of potato chips—both are acting within the
human sphere of experience of pleasures and pain. Temperance is the virtue in the human sphere of experiencing pleasure and pain that lies between
two extreme dispositions, which for Aristotle are the corresponding vices to
the virtue: licentiousness (excess of what is needed for being virtuous in this
domain) and insensibility (deficiency of what is needed).[47]

How could we understand the role of a disposition akin to temperance in
an ethic of sustainable well-being? Let me first start with its importance for
individual well-being. In the so-called developed countries, "material consumption" is an important way of satisfying a culturally specific version of
bodily appetites and pleasures—namely, the pleasure of getting new material
things.[48] Research on the link between materialism and well-being suggests,
for instance, that "the more materialistic values are at the center of our lives,
the more our quality of life is diminished."[49] This research provides some
empirical support for the important role of the disposition of temperance
with respect to feeling and acting upon desires for more material goods as
part of our life orientation in the sphere of pleasure and pain *for our individual well-being.*

The disposition of temperance is also important to the sustainability of the ecological system and the quality of life of future generations of humans—and thus to sustainable well-being—as research in the fields of sustainability,[50] environmental ethics,[51] and ecological economy[52] suggests. Thus, the disposition of temperance—in the sense illustrated here—plays a central role *as a means* in an ethic of sustainable well-being. Let us look at this point more closely, using the temperance example to illustrate the notion that virtues are defined by, rather than define, an ethic of sustainable well-being; as such, we cannot take for granted what we actually mean by "temperance" as a virtue in an ethic of sustainable well-being. This is because our understanding of individual well-being that is "supportive of" the sustainable development of social-ecological systems defines what makes a virtue in an ethic of sustainable well-being.

Duane Elgin and others have argued for "voluntary simplicity" in living our lives, whereby "simplicity" is understood and experienced not as sacrifice but rather as "a way of life that is outwardly simple and inwardly rich."[53] From the standpoint of an ethic of sustainable well-being, the disposition toward living with voluntary simplicity would be a good candidate for a virtue as a means toward living sustainably well. The virtue of "voluntary simplicity" is defined by an ethic of sustainable well-being, not the other way around. If the latter were the case, virtues would have to be defined independently of an ethic of sustainable well-being, in which case "voluntary simplicity" would most likely be considered not a virtue but rather a vice in Aristotle's scheme: from the standpoint of an Aristotelian ethics enacted in twenty-first-century North America, voluntary simplicity might be more likely considered a puritanical way of living and, thus, one of the two extreme dispositions within the domain of pleasure and pain (insensibility) rather than being the middle (virtuous) disposition of temperance in this domain of human living. However, from an ethic of sustainable well-being, we can say that such an ethic defines voluntary simplicity as a virtuous disposition in the sphere of pleasure and pain. If we want to keep Aristotle's system of virtues as the middle dispositions and the vices as the two extreme positions (excess and deficiency), we can say that in a normative ethic of sustainable well-being, the virtuous disposition in the sphere of pleasure and pain is voluntary simplicity, which would be the understanding of "temperance" in an ethic of sustainable well-being, while, for instance, the disposition toward living a life of consumption that has been normalized in Western societies[54] would be a vice in an ethic of sustainable well-being.

# Conclusion

Everyone has some understanding of what it means to be well because it is what we are striving for when exerting our agency as part of living our lives. This does not mean that our specific understanding is complete or explicit enough or even adequate in the assessment of our, or other people's, state of being or desired states of being. But it does mean that striving for our well-being is at the core of our lives' concerns, be they our day-to-day concerns or concerns at crucial junctures in our lives. As social beings, humans are concerned not only with their own well-being but also with the well-being of fellow humans. This concern is not only evident in our concern for the well-being of those close to us, but also in the communal concern at the political and educational level of communal living. Better understanding of and engagement with human well-being is, thus, core to our concern for our and our fellow humans' lives. From the bio-systemic starting point of the approach to human well-being, given that we only have our subjective experiences available to respond to, any concern for other living beings and, more generally, the earth's ecosystem can best be, and indeed should be, *integrated* right into our concern for human well-being. With future generations and the sustainable development of the earth's ecosystem in mind, this is a core contribution that I think this approach can make to the discourse on understanding human well-being and on assessing the quality of human living.

An integrated approach, as suggested in this chapter, requires much more space to develop than is available here. Hence, some of the ideas this chapter speaks about have to stay underdeveloped and under-rationalized, but I hope that what I was able to present here provides a cohesive overview of the main ideas and how they link together.

## Notes

1    Michael Borucke et al., "Accounting for Demand and Supply of the Biosphere's Regenerative Capacity: The National Footprint Accounts' Underlying Methodology and Framework [Draft]," *Global Footprint Network* (2013), 4, http://www.footprintnetwork.org/content/images/NFA%20Method%20 Paper%202011%20Submitted%20for%20Publication.pdf. More information of ecological footprint research can be found at www.footprintnetwork.org.

2   World Wildlife Fund, *Living Planet Report 2016: Risk and Resilience in a New Era* (Gland, CH: World Wildlife Fund International, 2016), 13, http://www.footprintnetwork.org/content/documents/2016_Living_Planet_Report_Lo.pdf.

3   Rashid Hassan, Robert Scholes, and Neville Ash, eds., *Ecosystems and Human Well-Being: Current State and Trends* (Washington, DC: Island Press, 2005).

4   For example, Thomas Piketty, *Capital in the Twenty-First Century* (Cambridge, MA: Belknap Press, 2014); Markus Stierli et al., *Global Wealth Report 2015* (Zurich: Credit Suisse AG, 2015).

5   For example, Raj Patel, *Stuffed and Starved: Markets, Power and the Hidden Battle for the World's Food System* (Toronto: Harper Collins, 2007).

6   For example, Juliet B. Schor, *Plenitude: The New Economics of True Wealth* (New York: Penguin, 2010), 61.

7   For example, Alan Drengson and Yuichi Inoue, eds., *The Deep Ecology Movement: An Introductory Anthology* (Berkeley, CA: North Atlantic Books, 1995); Arne Naess, *The Selected Works of Arne Naess* (10 vols.), ed. H. Glasser (Dordrecht, NL: Springer, 2005).

8   For example, Tim Jackson, *Prosperity without Growth: Economics for a Finite Planet* (London: Earthscan, 2009); Schor, *Plenitude.*

9   For example, Ronald L. Sandler, *Character and Environment: A Virtue-Oriented Approach to Environmental Ethics* (New York: Columbia University Press, 2009); Ronald Sandler and Philip Cafaro, eds., *Environmental Virtue Ethics* (Lanham, ML: Rowman & Littlefield, 2005).

10  K. Anthony Appiah, *Experiments in Ethics* (Cambridge, MA: Harvard University Press, 2008), 37.

11  Thomas Falkenberg, *Framing Human Well-being and Well-becoming: An Integrated Systems Approach* (Paper Series #2), 2018, http://wellbeinginschools.ca/wp-content/uploads/2019/09/WBIS-Paper-No-2-Falkenberg-2019-2.pdf.

12  This section draws on Falkenberg, *Framing Human Well-being and Well-becoming,* where the ideas presented here are developed far more extensively and in much further detail and with reference to the relevant literature. For reasons of space, I also do not elaborate here on the distinction between well-being and well-becoming.

13  Humberto R. Maturana and Francisco J. Varela, *Autopoiesis and Cognition: The Realization of the Living* (Dordrecht, NL: D. Reidel, 1980); Humberto R. Maturana and Francisco J. Varela, *The Tree of Knowledge: The Biological Roots of Human Understanding,* rev. ed. (Boston: Shambhala, 1998).

14  For example, Fritjof Capra, *The Web of Life: A New Scientific Understanding of Living Systems* (New York: Anchor Books, 1996); Fritjof Capra and Pier L. Luisi, *The Systems View of Life: A Unifying Vision* (Cambridge: Cambridge University Press, 2014).

15 Maturana and Varela, *The Tree of Knowledge*, 75.
16 Ernst von Glaserfeld, *Radical Constructivism: A Way of Knowing and Learning* (New York: Routledge Falmer, 1995), 18.
17 For a discussion of different approaches to well-being in the Western scholarly literature, see Thomas Falkenberg, "Making Sense of Western Approaches to Well-Being for an Educational Context," in *Sustainable Well-Being: Concepts, Issues, and Educational Practice*, ed. Frank Deer et al., 77–94 (Winnipeg: ESWB Press, University of Manitoba, 2014), and for Indigenous perspectives, see Frank Deer and Thomas Falkenberg, eds., *Indigenous Perspectives on Education for Well-Being in Canada* (Winnipeg: ESWB Press, University of Manitoba, 2016).
18 Falkenberg, "Making Sense of Western Approaches," 78–9.
19 Jack Martin, Jeff Sugarman, and Janny Thompson, *Psychology and the Question of Agency* (Albany: State University of New York Press, 2003).
20 Gordon Allport, "The Fruits of Eclecticism—Bitter or Sweet?" *Acta Psychologica* 23 (1964): 36.
21 For example, Martin E.P. Seligman and Mihaly Csikszentmihalyi, "Reply to Comments," *American Psychologist* 56, no. 1 (2001): 90.
22 John C. Christopher, "Situating Psychological Well-Being: Exploring the Cultural Roots of Its Theory and Research," *Journal of Counseling and Development* 77, no. 2 (1999): 149.
23 For ease of reading, I will henceforth shorten "well-being and well-becoming" to "well-being," except in the subheadings.
24 This conceptualization of social systems is an oversimplification. For more appropriate but also more complicated approaches to understanding social systems, see Gidden's structuration theory (e.g., Anthony Giddens, *The Constitution of Society: Outline of the Theory of Structuration* [Berkeley: University of California Press, 1984]) and Archer's morphogenetic approach (e.g., Margaret S. Archer, *Realist Social Theory: The Morphogenetic Approach* [Cambridge: Cambridge University Press, 1995]).
25 Maturana and Varela, *Autopoiesis and Cognition*; Maturana and Varela, *The Tree of Knowledge*.
26 Wade Davis, *The Wayfinders: Why Ancient Wisdom Matters in the Modern World* (Toronto: House of Anansi Press, 2009), 117.
27 Davis, *The Wayfinders*, 118–19.
28 Davis, *The Wayfinders*, 119.
29 For example, Fikret Berkes and Carl Folke (with Johan Colding), eds., *Linking Social and Ecological Systems: Management Practices and Social Mechanisms for Building Resilience* (Cambridge: Cambridge University Press, 1998); Lance H. Gunderson and Crawford Stanley Holling, eds., *Panarchy: Understanding Transformations in Human and Natural Systems* (Washington, DC: Island Press, 2002).

30 Crawford Stanley Holling, "Understanding the Complexity of Economic, Ecological, and Social Systems," *Ecosystems* 4 (2001): 399.

31 For example, Jon Norberg et al., "Diversity and Resilience of Social-Ecological Systems," in *Complexity Theory for a Sustainable Future*, ed. Jon Norberg and Graeme S. Cumming, 46–79 (New York: Columbia University Press, 2008).

32 Appiah, *Experiments in Ethics*, 37.

33 Karen Armstrong, *Twelve Steps to a Compassionate Life* (Toronto: Vintage Canada, 2011).

34 Greg Pence, "Virtue Theory," in *A Companion to Ethics*, ed. Peter Singer, 249–58 (Malden, MA: Blackwell, 1991).

35 Justin Oakley, "Varieties of Virtue Ethics," *Ratio* 9, no. 2 (1996): 133.

36 Oakley, "Varieties of Virtue Ethics," 133.

37 Oakley, "Varieties of Virtue Ethics," 134.

38 Oakley, "Varieties of Virtue Ethics," 133.

39 Oakley, "Varieties of Virtue Ethics," 133.

40 Oakley, "Varieties of Virtue Ethics," 133.

41 For example, Philippa Foot, *Natural Goodness* (Oxford: Clarendon Press, 2001).

42 Oakley, "Varieties of Virtue Ethics," 133, 134.

43 Rosalind Hursthouse, *On Virtue Ethics* (Oxford: Oxford University Press, 1999), 172.

44 Aristotle, *The Ethics of Aristotle: The Nicomachean Ethics*, trans. J.A.K. Thomson, rev. ed. (London: Penguin, 1976), bk. 2.

45 Martha Nussbaum, "Non-Relative Virtues: An Aristotelian Approach," in *The Quality of Life*, ed. Martha C. Nussbaum and Amartya Sen (Oxford: Clarendon Press, 1993), 248.

46 Aristotle, *The Ethics of Aristotle*, 104.

47 Aristotle, *The Ethics of Aristotle*, 104.

48 Jackson, *Prosperity without Growth*, ch. 6.

49 Tim Kassler, *The High Price of Materialism* (Cambridge, MA: MIT Press, 2002), 14. For more on this inverse relationship, see Jackson, *Prosperity without Growth*; Michael J. Sandel, *What Money Can't Buy: The Moral Limits of Markets* (New York: Farrar, Straus and Giroux, 2012); Schor, *Plenitude*; Ernst F. Schumacher, *Small Is Beautiful: Economics As If People Mattered* (New York: Harper & Row, 1973); Barry Schwartz, *The Costs of Living: How Market Freedom Erodes the Best Things in Life* (New York: W.W. Norton, 1994).

50 For example, Hassan, Scholes, and Ash, *Ecosystems and Human Well-Being*.

51 For example, Joshua C. Gambrel and Philip Cafaro, "The Virtue of Simplicity," *Journal of Agricultural and Environmental Ethics* 23, no. 1–2 (2010): 85–108; Sandler and Cafaro, *Environmental Virtue Ethics*.

52 For example, Jackson, *Prosperity without Growth*.

53  See, for instance, the recommended readings and resources listed in Mark A. Burch, *Stepping Lightly: Simplicity for People and the Planet* (Gabriola Island, BC: New Catalyst Books, 2011), and Duane Elgin, *Voluntary Simplicity: Toward a Way of Life that Is Outwardly Simple, Inwardly Rich*, 2nd ed. (New York: Harper, 2010).

54  Erik Assadourian, "The Rise and Fall of Consumer Cultures," in *State of the World 2010. Transforming Cultures: From Consumerism to Sustainability*, ed. Linda Starke and Lisa Mastny, 3–20 (New York: Norton & Company, 2010); Jackson, *Prosperity without Growth*; Schor, *Plenitude*.

# NINE

# Why Virtue Is Good for You

## The Politics of Ecological Eudaimonism

MIKE HANNIS

C LASSICAL GREEK VIRTUE ETHICS WAS BASED ON A EUDAIMONIST conception of virtue, meaning that the possession and exercise of the virtues was taken to benefit the possessor, by contributing to their own flourishing (*eudaimonia*). Indeed, this was precisely the criterion by which a character trait or disposition was judged to be a virtue at all. A good life was a virtuous life, and vice versa. As one summary has it, "Human well-being (*eudaimonia*) is the highest aim of moral thought and conduct, and the virtues (*aretê*: 'excellence') are the requisite skills and dispositions needed to attain it."[1]

For Plato, to live virtuously was to have a healthy soul whose rational, spirited, and appetitive parts are in harmonious relationship. Moreover, to live virtuously was also to live justly. To act unjustly was not only to fall short of what virtue required, but to have a mistaken conception of what one's own flourishing consisted of.[2] The broader political community also flourished as citizens developed and exercised virtues, the dispositions of character required for their own flourishing. Little or no conflict was envisaged between pursuing the good life for oneself and valuing the wider goods of others and of the community.

Of course, the awkward historical fact is that the society which produced this elegant ethical theory failed in practice to properly consider the flourishing of women, slaves, and other non-citizens. It might perhaps be argued that this fact alone suggests that eudaimonism is inadequate. Surely virtue must involve cultivating not only one's own flourishing, but also that of others, irrespective of their status? To a modern mind accustomed to concepts of universal human rights, and to distinguishing between political communities and moral ones, this seems a powerful objection.

But the eudaimonist has a reply. Concern for the good of others is indeed a central sphere for the exercise of virtue and for the avoidance of vice. The better a person does at exercising the relevant virtues (such as compassion and respect) and avoiding the relevant vices (such as selfishness and arrogance), the more he or she will flourish. A great deal of what is required for a flourishing human life concerns relationships with others, and cultivating good relationships requires treating others well. So the virtues whose development and exercise enhance one's own flourishing, properly construed, will also enhance the flourishing of others. Concern for the good of others does not mandate a non-eudaimonist conception of virtue.[3] This remains true when the others in question inhabit realms beyond the human.

## Ecological Eudaimonism

The field of environmental virtue ethics (EVE) has emerged more recently as an ethical response to anthropogenic ecological crisis. As in virtue ethics more broadly, there has been debate about whether EVE should be based on eudaimonism or should embrace other conceptions of virtue. Given the urgency of the problems EVE hopes to address, such debates may seem like an unwarranted digression into arcane meta-ethics. However, they have far-reaching implications, both theoretical and practical. In this chapter I defend a eudaimonist version of EVE on which ecological virtues are seen as virtues because their possession and exercise contribute to the flourishing of the human beings exercising them, rather than because of beneficial effects on non-humans or on the integrity of natural systems.

This does not, however, mean that such "external" benefits go unrecognized or unvalued. It would thus be a mistake to characterize ecological eudaimonism as "anthropocentric" in the commonly encountered pejorative sense. It does not assume some objectively superior place for humanity in the grand scheme of things, nor does it assert that "nature" is valueless except where it is instrumentally useful to humans as a provider of resources or services. Given a sufficiently rich and nuanced picture of human flourishing, which takes full account of the importance of interpersonal and ecological relationships, the concept of ecological virtue does not require the justification of character traits as virtues on the basis of external or non-eudaimonistic ends. I have argued elsewhere that this rich picture could potentially emerge from an ecologically literate version of a capabilities approach, and that capabilities and virtues can be seen as complementary parts of the substantive, more-than-procedural conception of the good life that is a prerequisite for effective and ethical environmental policy-making.[4]

On this view, the supposed cleavage between ecocentrism ("good") and anthropocentrism ("bad") is revealed as a false dichotomy that merely replicates the problematic separation of humans from "nature." For ecocentrists, an unwarranted anthropocentrism that takes account only of nature's value to humans lies at the root of our ecological crisis: only by moving to an ecocentric perspective can we recognize the true intrinsic value of the non-human world, and from there proceed to discover the proper ways to interact with it. Our own flourishing—or as deep ecologists (e.g., Arne Naess)[5] put it, our Self-realization—depends in turn on this move away from anthropocentrism. But the critical question facing us as contemporary human beings is not the abstract one of whether "nature has value for itself."[6] It is the very practical one of whether we can find ways to flourish in symbiosis with the wider world, rather than as parasites upon it.

Thomas Hill observed in 1983 that cases in environmental ethics often evoke a "moral uneasiness" that is not adequately captured either by utilitarian arguments about balancing competing interests and maximizing utility, nor by deontological ones about rights and duties.[7] Rather, it becomes necessary to ask questions like "*What kind of person ... would cover his garden with asphalt, strip-mine a wooded mountain, or level an irreplaceable redwood grove?*"[8]

Hill's answer is that such a person typically lacks "that sort and degree of humility that is a morally admirable character trait."[9] In more overtly

Aristotelian style, Geoffrey Frasz posits a virtue of "openness," defined as a mean between the opposing vices of false modesty and arrogance. A person possessing this virtue, he suggests, "is neither someone who is closed off to the humbling effects of nature nor someone who has lost all sense of individuality when confronted with the vastness and sublimity of nature."[10] John Barry argues that a person "closed off to the humbling effects of nature" exhibits what Ehrenfeld famously called "the arrogance of humanism."[11] For Barry, however, ecocentrism is not the answer, since the vice to be avoided is arrogance rather than humanism. In fact he identifies ecocentric positions with the opposite extreme, seeing "unreflective sentimental or romantic views of human-nature relations" as an ecological vice, equivalent to Frasz's "loss of individuality." The key ecological virtue of humility (equating roughly to what Frasz calls openness) is thus insightfully described by Barry as "a mean between a timid ecocentrism and an arrogant anthropocentrism."[12]

These writers understand the possession and exercise of the relevant virtue(s) not only to benefit other people and the wider natural world but also to be important for the flourishing of the individual. As Barry points out, this is congruent with Alasdair MacIntyre's view of the good life for human beings, in which the "virtues of acknowledged dependence" are critically important for the flourishing of the individual, as well as for the integrity of the community and the "networks of giving and receiving" of which that individual is a part.[13] This insight may usefully be extended beyond the human community, to consider as virtues of acknowledged *ecological* dependence those character traits and dispositions that tend to maintain the integrity of the broader natural world.[14] As with virtues of acknowledged inter-human dependence, possession and exercise of these traits and dispositions benefits the individual, not only because such integrity is itself important for human flourishing but also because recognition and acknowledgement of our dependencies and vulnerabilities is vital to developing a proper understanding of our own identity.

## Virtue, Respect, and Orientation

Ronald Sandler makes a distinction between ecological virtues of prudence and ecological virtues of respect, arguing that the former contribute to the flourishing of the human agent, and the latter to the flourishing of non-humans.

Noting that "environmental virtues and vices are legion," he lists twenty-seven relevant virtues, subdivided into six categories.[15] For Sandler, the "virtues of sustainability" (temperance, frugality, farsightedness, attunement, and humility) are justified in prudential, eudaimonist terms. The same is true of his "virtues of communion with nature," which include receptivity, wonder, love, and aesthetic sensibility.[16] While the possession and exercise of all these virtues is likely to have beneficial effects for non-humans and be effective in preserving the integrity of ecosystems, it also contributes to human flourishing, and it is this contribution that makes them virtues. By contrast, his "virtues of respect for nature" (ecological sensitivity, care, compassion, restitutive justice, and non-maleficence) are justified in non-eudaimonist terms. That is, they are seen as virtuous because of their contribution to the flourishing of non-human entities, *not* because of any direct contribution to that of humans.[17] Sandler's is therefore a hybrid or "pluralist" theory of virtue, embracing both eudaimonist and non-eudaimonist conceptions.

There has been substantial debate about this hybrid structure, reflecting more general differences of opinion between virtue ethicists about eudaimonism.[18] I discuss below some political and practical implications of such debates for the promotion of ecological virtue. First, however, it is important to note the connection between the justification of an ecological virtue in non-eudaimonist terms and the implied intrinsic value of the end taken to justify it, such as the flourishing of a non-human individual or the integrity of an ecosystem.

My own view is that EVE need not be based on the controversial notion of intrinsic value in the non-human world, and that given the controversy, it will be more robust if it is not. With sufficient attention to the value of the *relationships* between humans and the non-human world, it seems likely that all the virtues on Sandler's list (and others—see, for example, Santas[19]) could instead be justified in eudaimonist terms as contributing to human flourishing. Following Rosalind Hursthouse,[20] I argue that ecological virtues are better justified in such eudaimonist terms. Taking the influential work of Paul Taylor as a departure point, Hursthouse argues that respect for nature should be seen as "a character trait rather than an attitude."[21] Respect for nature cannot simply be adopted, but requires education and practice:

> What [Taylor] describes … is being rightly oriented to nature, through and through, in action, emotion, perception, sensibility and understanding.

What is involved in "adopting" this attitude would ... manifestly have to be a complete transformation of character. ... Can having "respect for nature," as he describes it, not come about at all, given that it cannot simply be adopted or taken up? The problem is solved if we construe it as a virtue. You can't just decide to have ... virtuous character traits. But they can be acquired through moral habituation or training, beginning in childhood and continued through self-improvement.[22]

Taylor's ethical rationale for respecting nature rests on the attribution of "inherent worth" to living things.[23] The fact that living things have a *telos*, or "a good of their own," means that they are "members of the Earth's Community of Life."[24] This in turn means that, like humans, they are entitled to respect, in an explicitly Kantian sense of entitlement to always being treated as ends rather than only as mere means. Thus, also like humans, they should be treated as having intrinsic value. (This is, broadly speaking, also the reasoning behind Sandler's non-eudaimonist justification of his more fine-grained "virtues of respect for nature.")

At first sight this all seems reasonable. Yet a number of problems follow from building in intrinsic value as a foundational premise in this way. Some are meta-ethical problems, such as what exactly it would mean for value to exist independently of valuers. Others are more practical, such as the thorny question of resolving competing claims between the interests of living things, once both are considered to have intrinsic value. Do some living things—for instance humans, or perhaps whales—have *more* intrinsic value than others? Neither yes nor no seem satisfactory answers. There is something deeply wrong with the question, not least because even to ask it is to beg other equally problematic questions, such as, "Is intrinsic value quantifiable?"

As Hursthouse points out, a further shortcoming of Taylor's "biocentric deontology" is that it struggles to ground respect for *inanimate* natural features or systems. His "community of life" excludes the "soils and waters" that are explicitly included in Aldo Leopold's celebrated "expanded community" (even though Leopold confusingly refers to this community as "biotic"[25]) and seems capable of justifying respect for the integrity of natural systems only indirectly, via respect for their living components. The "deep ecology" response to this shortcoming is to move from Taylor's biocentrism to a fully-fledged ecocentrism, extending ideas of teleology and intrinsic value further still to encompass mountains, oceans, ecosystems, and so forth. Hursthouse argues that a virtue approach renders this problematic expansion unnecessary, since from a virtue ethics perspective, intrinsic value is best understood as a colloquial or heuristic concept, not a foundational theoretical commitment. As such it is not required for the normative justification of respect for nature. Perhaps surprisingly, then, moving away from intrinsic value can *widen* the scope of natural entities, systems, and phenomena for which respect is justified. As Hursthouse observes,

> If we think of being "rightly oriented to nature" not as an attitude founded on an adult's rational recognition of [the intrinsic value of living things] but as a character trait arising from a childhood training that gives us particular reasons for action (and omission) in particular contexts, and shapes our emotional response of wonder, the hard and fast line [Taylor] draws between the animate and the inanimate becomes insignificant.[26]

This notion of right orientation to nature, understood as a virtue, provides a way of explicitly considering the kinds of relationships with the non-human to which an individual (or indeed a culture) is disposed. Environmental virtue ethics will be better equipped to conceptualize and guide such relationships if it is unhampered by a monistic focus on the recognition and measurement of intrinsic value. After all, a key attraction of a virtue approach is that it allows for flexibility and recognizes that each new context is likely to be different. Each fresh problem requires the relevant virtue(s) to be applied in fresh ways appropriate to the unique context, applying the "meta-virtue" of judgment to work out just what these ways might be. As Brian Treanor notes:

> Virtue ethics in the Aristotelian tradition is fully committed to the uniqueness, multiplicity and noncommensurability of ethical goods. ... On the virtue ethicist's account we cannot simply compare ethical goods "scientifically" in terms of some single metric that we would then seek to maximise, as is the case with utilitarianism.[27]

Uniqueness, multiplicity, and incommensurability are of course key features not only of the ethical landscape but of physical landscapes too. No universal formula can dictate the best course of action in every case of human impact on the richly multi-faceted non-human world. Scientific analysis and

mathematical calculation are essential tools, but they can never replace qualities such as wisdom, judgment, or "right orientation." Environmental ethics
must stand firm against the increasing tendency in environmental economics
to reduce the wonderful complexity of the natural world to universal metrics
and formulae that can be applied with little or no explicit ethical reflection
through techniques such as natural capital accounting, biodiversity offsetting,
and payments for ecosystem services.[28] In its attempt to standardize the exuberant variety of the real world, such thinking always struggles with incommensurability.[29] In ethical terms, this kind of balance-sheet thinking seems like
utilitarian reasoning *par excellence*: yet even in these terms it fails, because it
inevitably disregards important factors that cannot be meaningfully quantified
and fitted into models.[30]

## Absent Exemplars?

Hursthouse is pessimistic about the prospects of knowing what it means in
practice to be rightly oriented to nature, suggesting that "it is possible that we
have already made such a mess that we shall not be able to live well, as part of
the natural world, for many generations to come, if ever."[31] Part of this pessimism comes from her conviction that we lack exemplars or *phronimoi*, wise
people from whom to learn *what it is* to live in such a way:

> It is possible, though this is contested, that we have glimpses of what it might
> have been like to live in accordance with the virtue of being rightly oriented
> to nature in the little we know of the lives of the Australian Aborigines and
> the Amerindians before European hegemony. But even if we knew a lot
> more about their lives and even if it were certain that they had possessed
> the virtue, this would not entail that that is how we should strive to live
> and be now. Human beings are essentially socially and historically situated
> beings and their virtuous character traits have to be situated likewise. A
> twenty-first-century city-dweller who possessed the virtue to some degree
> could hardly manifest it in just the same ways as Australian Aborigines and
> Amerindians perhaps used to when they lived as hunter-gatherers. What we
> need to know is what would count as living in accordance with it *now* or in
> the near future.[32]

The problem of how best to act within a society that is itself operating unethically in some important respect is not a new one. Modern representative democracy tries, but usually fails, to offer a political answer to this long-standing ethical problem. Electing a significantly different government may not be possible, and indeed may not have any impact on the problem at hand: it may, for instance, be the case that a majority supports unjust arrangements. In such circumstances law-breaking and direct action may well be justified. But these will still need to be based on coherent ethical principles. On the face of it no existing ethical tradition seems perfectly equipped to provide ethical guidance that does not rely on the background assumption that the person needing guidance lives in a basically just and well-functioning society.[33] Nonetheless, virtue ethics, with its focus on the agent, can perhaps adapt more easily to this difficult task than other frameworks. As Hursthouse identifies, though, any such adaptation does require an answer to the problem of absent *phronimoi*.

Brian Treanor points out that narrative, whether historical or fictional, can provide "virtual *phronimoi*," and thereby "throw a bridge—the bridge of as-if experience—between abstract, faceless, lifeless theory and concrete, particular, lived experience." This is unsurprising, he argues, since "our ideas about flourishing—happiness, success, excellence, what constitutes a worthwhile life, and so on, are primarily grasped and articulated in narratives."[34] Viewed in this light, historical knowledge about life "before European hegemony" may well contain valuable and relevant lessons about what it is to be "rightly orientated to nature." Also helpful here is Philip Cafaro's suggestion that acting in such a way as to avoid the all-too-widespread examples of ecological vice may help us emulate more positive exemplars from our own cultures.[35]

Still, looking for living exemplars of the virtue in other human cultures is a worthwhile endeavour, and Hursthouse is far from alone in recommending it. Her pessimism about actually *finding* any such exemplars is, however, overstated. It is still possible to learn about different ways of relating to the non-human world from people who have not yet been coerced or conditioned into modern consumerist relationships with "nature," and there is no good reason to assume that modern societies cannot learn anything useful from the *phronesis* (practical wisdom) of such cultures. Learning about different orientations toward nature is not a matter of copying specific practices, or of seeking out "unspoiled" native inhabitants of mythically pristine environments.

Consider, for instance, the San peoples of the Kalahari, influentially roman-
ticized by Laurens van der Post[36] and many others. More attentive scholarship
has made clear that far from being magically intact relics of a "pre-lapsarian"
Stone Age, these are in fact people with a long history of being forced off more
fertile land by more aggressive pastoralists and agriculturalists.[37] The realities of
their present precarious existence make clear that it has been—and remains—
not only misleading but damaging to represent them as "closer to nature" in
some Rousseauvian "noble savage" way. Yet this does not mean there is nothing
to learn from them about human relationships with the non-human world.

Building on Richard Lee's meticulous work with San people, Marshall
Sahlins famously developed a profound and provocative reinterpretation
of "affluence" as the ability to meet one's needs from available resources.
Notwithstanding their limited resources, he argued, well-functioning foraging
societies historically enjoyed a high quality of life and indeed had considerably
more "leisure time" than workers in modern industrialized societies.[38] Key to
this are forms of social organization characterized by a very strong emphasis
on sharing and reciprocity. This "primitive communism"[39] not only resulted in
everyone getting food to eat but also, more broadly, meant that the economic
or market "scarcity" so taken for granted in modernity was entirely absent.

By demonstrating the possibility of aligning ends with means, cases like that
of the San thus show that far from being a fundamental feature of the human
condition, the assumption of *unlimited* ends so characteristic of modern economics, and of associated hegemonic pictures of "human nature," is contingent and culturally determined. It should be noted that vigorous disagreement
continues among anthropologists about these matters. Still, Nurit Bird-David's
reinterpretation of Sahlins, as summarized by Jacqueline Solway, strongly suggests that at least for some, MacIntyre's "networks of giving and receiving" are
understood to extend well beyond the human:

> Hunter-gatherers achieve abundance not through simple trust or confi-
> dence in the environmental bounty that is "out there" and ontologically
> distinct from themselves, but rather the environment is assimilated into a
> larger world of "giving and taking" in which foragers themselves are among
> its many members. Foragers' confidence resides in that "world" and the fact
> that their needs will be met through its networks of sharing, in which they
> are enmeshed.[40]

Clearly, then, egalitarianism is one thing modern societies can learn about from other cultures, both past and present.[41] Anthropologists have worked hard to understand how existing and historical (as opposed to theoretical) egalitarian societies remain so in the face of formidable pressures, debating for instance whether or not equality can be a stable condition, and considering how such lessons might be applied to modern contexts.[42] Lessons about egalitarianism might not immediately seem to equate to lessons about "right orientation to nature." But as discussed further below, egalitarianism (or the lack of it) is a crucial determinant of human well-being, of consumption patterns, and of ecological sustainability.

More broadly, the historical and cross-cultural perspectives offered by anthropology provide a salutary reminder that the modern capitalist nation-state has largely superseded other forms of social organization not because of some evolutionary superiority, but due to a combination of historical contingency and brute force.[43] This violent transition has, of course, coincided with the well-documented ecological devastation that forms the background to current speculation about ecological virtue. Should it be confirmed that the relationship between these two phenomena is causal rather than coincidental, it will become even more important to study whatever knowledge remains of other cultures' attitudes to the non-human world.

## Eudaimonist Environmental Policy

Writers on environmental or ecological citizenship routinely note that changing environmental attitudes and behaviour is more reliably achieved by encouraging people to re-examine their values than by the "carrot-and-stick" route of financial incentives and penalties.[44] Building a widespread sense of citizenship, including environmental citizenship, is seen not only as ethically preferable but also as more effective than appeals to self-interest.

Consideration of free-rider problems and issues of delayed effect leads to the same conclusion. For instance, in the salient case of carbon emissions, there is a delay of at least twenty to thirty years between emission and impact.[45] Any climate-related benefit accruing from foregoing a flight today will not be felt for decades. Hence it is by no means an "irrational" choice for a self-interested *homo economicus* to live it up now, leaving future generations to pay the price.[46] Nor would it be irrational to calculatedly continue reaping the material benefits of a high-emission lifestyle while enjoying the environmental benefits of others' reductions. Appealing purely to self-interest, it seems, will never deliver the required changes in beliefs and behaviours.

When dealing with real people rather than with a mythical *homo economicus*, however, there is (perhaps ironically) a sense in which the concept of ecological virtue makes appealing to self-interest seem viable after all. The exercise of the virtues, on a eudaimonist model, makes for a genuinely better life. If I believe that an ecologically virtuous life will be a happier or more flourishing one, I may be motivated to change my behaviour accordingly. This suggests a powerful political strategy. The ecologically virtuous life, if it is likely to be a better one, can truthfully be presented as its own reward.

There are practical as well as ethical limits to what can be achieved by a regulatory enforcement approach to environmental policy. It would also be unrealistic to expect altruistic motivations to suddenly, and unprecedentedly, overwhelm self-interest in consumerist "liberal democratic" societies. Fortunately, however, humanity does not face the stark choice between accepting coercion and hoping for altruism. The promotion and facilitation of ecological virtue can build on a sincere appeal to genuine self-interest, based on an ecologically informed notion of human flourishing.

It can also do important work in mitigating the very real problem of uncertainty. The devalued academic reputation of the *Limits to Growth* report[47] vividly illustrates the dangers of modelling using empirical estimates of ecological limits, technological trajectories, prices, and population levels.[48] All estimates and assumptions, however "state of the art," become hostages to fortune as history unfolds, and there will always be unavoidable uncertainties about precisely where limits should be set. This empirical uncertainty is sometimes used (as by the current US administration) to support the specious argument that it is unnecessary or unjust to allow ecological limits to influence economic policy at all. Ecological eudaimonism, by contrast, encourages us to celebrate human ecological embeddedness rather than vainly seeking to transcend it. The credibility and legitimacy of appropriate regulatory limits can be enhanced by appealing to ecological virtue as a justification for adopting a precautionary approach, rather than always pushing to the (inevitably unclear) limit.

But a caveat is critical here. If directed only at individuals, appeals to ecological virtue will be ineffective as well as disingenuous. For a public authority to adopt the rhetoric without genuinely and sincerely reorientating environmental, economic, and social policy to *embody and facilitate* such virtue would add nothing to existing ineffectual exhortations to good environmental citizenship. It would be both inconsistent and ethically objectionable, and thus very unlikely to gain support from anyone seeking more from environmental politics than a "performance of seriousness."[49] Objecting to hypocrisy is different from, and arguably more universally legitimate than, objecting to paternalism: if internalization of particular values is to be expected of citizens, then social and economic policy must itself clearly embody those values. For instance, exhorting individuals to cut their carbon emissions by flying less, while simultaneously building new airports, is likely to fail, but more because the exhortation is hypocritical than because it is paternalistic.

Furthermore, ecological virtue cannot be credibly promoted without concurrent action to remove impediments to such virtue. Many such impediments are more accurately seen as incentives to ecological vice. These incentives may be offered deliberately or accidentally, by government or by the private sector. Ecological virtue, and hence its promotion, serves human flourishing by (among other benefits) helping to bring about life-enhancing experiences of wholeness, relatedness, and coherence. Ecological vice endangers human flourishing, not only through the direct material consequences of environmental degradation but also in more subtle ways, such as the emergence of mental health problems related to the lack or loss of such experiences.[50] Public authorities should not be in the business of simply exhorting individuals to reconsider their values and become ecologically virtuous. The task at hand is a harder but more worthwhile one: to reconfigure social and economic conditions to *facilitate* ecological virtue.

## Facilitating Ecological Virtue

It is uncontroversial that reducing unsustainable levels of production and consumption can assist with the transition from our current predicament to a world of ecologically sustainable human societies, and that this transition would serve the interests both of future people and of non-humans. The controversial question is whether the measures required for the transition can truly be said to serve the interests of *present* people. Ecological eudaimonism answers this question in the affirmative. Coupled with an ecologically informed understanding of the capabilities required for flourishing human lives, it makes clear that it *is* in everyone's interest now to, in John Barry's phrase, "cultivate modes of character and acting in the world which encourage social-environmental relations which are symbiotic rather than parasitic."[51]

This means that ecological virtue need not be presented as a matter of prioritizing citizenly duty over one's personal interests. This seems strategically important for environmentalism generally, and particularly for notions of environmental citizenship. Whether the relevant community is conceived as a conventional (local or cosmopolitan) human polis, as in most environmental citizenship literature,[52] or as a Leopoldian "community" reaching beyond the human[53] is irrelevant here. In either case, from the perspective of ecological eudaimonism, exhortations to "good environmental citizenship" that present ecological virtue or "pro-environmental behaviour" as a matter of worthy sacrifice, in which the greater good is prioritized above one's own self-interest, rest on an underdeveloped notion of self-interest.

Anyone making such exhortations puts themselves in the strategically disadvantageous position of attempting to persuade people to act against their own interests. This perspective throws new light on the claim that environmentalists need to overtly discuss and promote underlying values, rather than tailoring their message to appeal to individuals' self-interest as economically rational consumers.[54] This is an important insight, yet as discussed above there is a sense in which the appeal to self-interest could usefully be broadened rather than abandoned. (Perhaps an appeal to a Naessian "Self-interest"—with a capital S—could replace appeals to *homo economicus*.) An ecologically informed conception of human flourishing, and the understandings of virtue that flow from this, could ground a fuller and more relational conception of self-interest entirely compatible with, and indeed served by, a radical shift toward more ecologically sustainable lifestyles and social structures.

As noted by Treanor, nurturing ecological virtue requires changing narratives. The currently dominant narratives about consumption are those unashamedly seeking to increase it. Estimates of the number of commercial advertising messages people in the United Kingdom or the United States are exposed to each day range from 250 to well over 3,000.[55] Commercial advertising deliberately creates "aspiration treadmills" and constitutes a "perverse incentive for unsustainable status competition."[56] Wilkinson and Pickett note that, feeding off both actual and perceived inequalities, advertising "amplifies and makes use of vulnerabilities which were there anyway" to drive "competitive consumption," frustrating efforts to reduce ecological impacts.[57] In virtue language, then, commercial advertising clearly constitutes a formidable incentive to ecological vice, which if left unaddressed is likely to undermine any efforts to facilitate ecological virtue. These ubiquitous messages dominate not only by sheer weight of numbers. Like other effective forms of deliberate psychological manipulation, the reason they are so effective is that they fundamentally distort how people relate to the world.

But unlike some others, this version of the widespread problem of "adaptive preferences" has potential solutions. Outlawing such manipulation is probably a precondition for any broader project aimed at circumscribing levels of production and consumption. In fact, curtailing the ceaseless upward manipulation of material expectations would probably represent the single most efficient way to intervene in the destructive cycle of overproduction, overconsumption, and ecological damage. It might also be the most effective climate change mitigation measure immediately available. Moreover, empirical research makes clear that there is no significant downside here. Above minimum thresholds, neither increased consumption nor increased "consumer choice" increases life satisfaction, because neither contributes significantly to a flourishing life.[58]

Moving from consumption to production, a society characterized by ecological virtue would be one in which people felt both less compelled and less willing to fill jobs that involve ecologically vicious actions. Achieving this would require not only changing management priorities but also addressing the disempowerment of those much lower down in corporate pecking orders, whose jobs consist of doing unsustainable things ordered by others. Employment should never have to mean leaving your ethics at home. A person who can only survive or feed their family by remaining in a job that involves doing things they feel to be wrong is clearly not flourishing. Yet this is a common situation all over the world, in richer nations as much as in poorer ones. One partial remedy is suggested by research showing that enterprises owned and run by those who work in them, rather than to provide financial returns for external investors, are both more able and more likely to be guided by ethical considerations, in environmental matters as in others.[59]

# Conclusion

Facilitating flourishing lives is the legitimate aim of government, and effective policies directed at ecological sustainability can and should be viewed as a key part of achieving this aim.[60] A sustainable society is one that effectively safeguards future people's capabilities for flourishing in an ecologically intact world. Ecological virtue contributes to individuals' flourishing, and understanding it in this eudaimonist way makes clear that environmental policy can and should include the sincere promotion and facilitation of ecological virtue, alongside eradication of incentives to ecological vice. Within this framework, egalitarianism may be considered a key ecological virtue.

## Acknowledgements

Completion of this chapter has been generously facilitated by funding from the UK Arts and Humanities Research Council under the Future Pasts project (AH/K005871/2: www.futurepasts.net) led by Sian Sullivan, to whom I am much indebted.

## Notes

1   Dorothea Frede, "Plato's Ethics: An Overview," *Stanford Encyclopedia of Philosophy*, 2016, https://plato.stanford.edu/archives/win2016/entries/plato-ethics/.

2   Aristotle largely agreed, albeit with some important differences: see Mark LeBar and Michael Slote, "Justice as a Virtue," *Stanford Encyclopedia of Philosophy*, 2016, https://plato.stanford.edu/archives/spr2016/entries/justice-virtue/.

3   The exclusionary prejudices of ancient Athens are of course in no way excused by this argument. But the problem lies with the prejudices, and the resulting narrow conception of the moral community, rather than with eudaimonism.

4   Mike Hannis, *Freedom and Environment: Autonomy, Human Flourishing, and the Political Philosophy of Sustainability* (New York: Routledge, 2016). See also John O'Neill, *Ecology, Policy and Politics* (London: Routledge, 1993), and Martha Nussbaum, *Creating Capabilities* (Cambridge, MA: Harvard/Belknap, 2011).

5   Arne Naess, *Ecology, Community and Lifestyle* (Cambridge: Cambridge University Press, 1989).

6   Haydn Washington et al., *Statement of Commitment to Ecocentrism*, 2017, http://
    www.ecologicalcitizen.net/statement-of-ecocentrism.php.

7   Thomas Hill, "Ideals of Human Excellence and Preserving Natural Environments,"
    in *Environmental Virtue Ethics*, ed. Ronald Sandler and Philip Cafaro (Lanham,
    MD: Rowman and Littlefield, 2005), 49.

8   Hill, "Ideals of Human Excellence," 50, emphasis added.

9   Hill, "Ideals of Human Excellence," 59.

10  Geoffrey B. Frasz, "Environmental Virtue Ethics: A New Direction for
    Environmental Ethics," *Environmental Ethics* 15 (1993): 279.

11  David Ehrenfeld, *The Arrogance of Humanism* (Oxford: Oxford University Press, 1978).

12  John Barry, *Rethinking Green Politics* (London: Sage, 1999), 31–6.

13  Alasdair MacIntyre, *Dependent Rational Animals* (London: Duckworth, 1999).

14  Mike Hannis, "The Virtues of Acknowledged Ecological Dependence,"
    *Environmental Values* 24, no. 2 (2015): 145–64.

15  Ronald Sandler, *Character and the Environment* (New York: Columbia University
    Press, 2007), 145.

16  Sandler, *Character and the Environment*, 82–3.

17  Sandler, *Character and the Environment*, 72.

18  For an overview of the disputed role of eudaimonism in virtue ethics more broadly,
    see William Prior, "Eudaimonism and Virtue," *Journal of Value Enquiry* 35, no. 3
    (2001): 325–42. For application of this debate to Sandler's work, see the exchange
    between Sandler, McShane, and Thompson in *Ethics, Place and Environment* 11, no.
    2 (2008): Ronald Sandler, "Natural Goodness, Natural Value, and Environmental
    Virtue: Responses to Katie McShane and Allen Thompson," 226–35; Allen
    Thompson, "Natural Goodness and Abandoning the Economy of Value: Ron
    Sandler's Character and Environment," 218–26; Katie McShane, "Virtue and
    Respect for Nature: Ronald Sandler's Character and Environment," 213–18.

19  Aristotelis Santas, "Aristotelian Ethics and Biophilia," *Ethics, Place and
    Environment* 19, no. 1 (2014): 95–121.

20  Rosalind Hursthouse, "Environmental Virtue Ethics," in *Working Virtue*, ed.
    Rebecca Walker and Phillip Ivanhoe, 155–71 (Oxford: Oxford University
    Press, 2007). See also Rosalind Hursthouse, *On Virtue Ethics* (Oxford: Oxford
    University Press, 1999).

21  Hursthouse, "Environmental Virtue Ethics," 162; Paul W. Taylor, *Respect For
    Nature* (Princeton: Princeton University Press, 1986).

22  Hursthouse, "Environmental Virtue Ethics," 163–4.

23  Taylor's "inherent worth" is broadly equivalent here to what is more commonly
    called intrinsic value. See John O'Neill, "Varieties of Intrinsic Value," *The Monist* 75
    (1992): 119–37; Piers Stephens, "Nature, Purity, Ontology," *Environmental Values* 9
    (2000): 267–94.

24 Taylor, *Respect For Nature*, 44.

25 Aldo Leopold, *A Sand County Almanac* (1949; repr., Oxford: Oxford University Press, 1968).

26 Hursthouse, "Environmental Virtue Ethics," 166.

27 Brian Treanor, *Emplotting Virtue: A Narrative Approach to Environmental Virtue Ethics* (New York: SUNY Press, 2014), 126–7.

28 Sian Sullivan, "On 'Natural Capital,' 'Fairy-Tales' and Ideology," *Development and Change* 48, no. 2 (2017): 397–423; Sian Sullivan and Mike Hannis, "Nets and Frames, Losses and Gains: Value Struggles in Engagements With Biodiversity Offsetting Policy in England," *Ecosystem Services* 15 (2015): 163–72; Sian Sullivan and Mike Hannis, " 'Mathematics Maybe, But Not Money': On Balance Sheets, Numbers and Nature in Ecological Accounting," *Accounting, Auditing & Accountability Journal* 30, no. 7 (2017): 1459–80.

29 Hannis, *Freedom and Environment*.

30 Mike Hannis, "Killing Nature to Save It? Ethics, Economics and Rhino Hunting in Namibia," *Future Pasts Working Paper* no. 4 (2016), https://www.futurepasts.net/fpwp4-hannis-2016.

31 Hursthouse, "Environmental Virtue Ethics," 170.

32 Hursthouse, "Environmental Virtue Ethics," 169.

33 Jeremy Bentham wanted utilitarianism to do exactly this—but evaluations of pleasure and pain are themselves culturally determined. In the context of "valuing nature," utilitarian approaches, notwithstanding their dominance, are also fatally susceptible to problems of incommensurability.

34 Treanor, *Emplotting Virtue*, 50, 120, 175.

35 Philip Cafaro, "Gluttony, Arrogance, Greed and Apathy: An Exploration of Environmental Vice," in *Environmental Virtue Ethics*, ed. Ronald Sandler and Philip Cafaro, 135–58 (Lanham, MD: Rowman and Littlefield, 2005); Philip Cafaro, "Thoreau, Leopold and Carson: Towards an Environmental Virtue Ethics" *Environmental Ethics* 23, no. 1 (2001): 3–17. See also John Barry, *The Politics of Actually Existing Unsustainability* (Oxford: Oxford University Press, 2012).

36 Laurens van der Post, *The Lost World of the Kalahari* (London: Vintage Books, 1958).

37 See, for example, James Suzman, *Affluence without Abundance: The Disappearing World of the Bushmen* (New York: Bloomsbury, 2017).

38 Marshall Sahlins, *Stone Age Economics* (Chicago: Aldine, 1972).

39 Richard B. Lee, "Demystifying Primitive Communism," in *Civilization in Crisis: Anthropological Perspectives*, ed. Christine W. Gailey, 73–94 (Gainesville: University Press of Florida, 1992).

40 Jacqueline Solway, " 'The Original Affluent Society': Four Decades On…," in *The Politics of Egalitarianism: Theory and Practice*, ed. Jacqueline Solway (New York: Berghahn Books, 2006), 71–2.

41 Mike Hannis and Sian Sullivan, "Reciprocity and Flourishing in an African Landscape," in *That All May Flourish: Comparative Religious Environmental Ethics,* ed. Laura Hartmann, 279–96 (Oxford: Oxford University Press, 2018).

42 See, for example, Bruce G. Trigger, "All People Are (Not) Good," in *The Politics of Egalitarianism: Theory and Practice,* ed. Jacqueline Solway (New York: Berghahn Books, 2006), 21–30. It is striking that the substantial literature on egalitarianism in anthropology remains almost entirely separate from that in political philosophy.

43 See, for example, Harold Barclay, *People without Government: An Anthropology of Anarchy,* 2nd ed. (London: Kahn & Averill, 1990); James C. Scott, *Seeing Like a State: How Certain Schemes to Improve the Human Condition Have Failed* (New Haven, CT: Yale University Press, 1998).

44 Andrew Dobson, *Citizenship and the Environment* (Oxford: Oxford University Press, 2003); James Connelly, "The Virtues of Environmental Citizenship," in *2006 Environmental Citizenship,* ed. Andrew Dobson and Derek Bell (Cambridge: MIT Press, 2006), 49–74; Mark Sagoff, *The Economy of the Earth,* 2nd ed. (New York: Cambridge University Press, 2008); Mark J. Smith and Piya Pangsapa, *Environment and Citizenship* (London: Zed Books, 2008).

45 IPCC (Intergovernmental Panel on Climate Change), *Fifth Assessment Report,* United Nations, 2014.

46 The interpretation of rationality as economic maximization is of course highly contested. For ecologically informed alternatives, see, for example, John Dryzek, *Rational Ecology: Environment and Political Economy* (Oxford: Blackwell, 1987); Val Plumwood, "Inequality, Ecojustice and Rationality," in *Debating the Earth,* ed. John Dryzek and David Schlosberg, 608–32 (Oxford: Oxford University Press, 1999).

47 Dennis H. Meadows et al., *Limits to Growth* (New York: Universe Books, 1972).

48 Detailed analysis has nonetheless found that Meadows's "standard run" scenario, which results in the collapse of the global system midway through the twenty-first century, has so far proved remarkably accurate. Graham Turner, "Is Global Collapse Imminent? An Updated Comparison of *The Limits to Growth* with Historical Data," MSSI Research Paper 4, University of Melbourne, 2014.

49 Ingolfur Blühdorn, "Sustaining the Unsustainable: Symbolic Politics and the Politics of Simulation," *Environmental Politics* 16, no. 2 (2007): 251–75.

50 Richard Louv, *Last Child in the Woods: Saving Our Children from Nature-Deficit Disorder* (Chapel Hill, NC: Algonquin Books, 2005); Glenn Albrecht, " 'Solastalgia': A New Concept in Health and Identity," *Philosophy, Activism, Nature* 3 (2005): 41–55; Theodore Roszak, Mary Gomes, and Allen Kanner, *Ecopsychology* (San Francisco: Sierra Club, 1995).

51 Barry, *Rethinking Green Politics,* 35.

52 See, for instance, Smith and Pangsapa, *Environment and Citizenship,* and contributions to Dobson and Bell, *2006 Environmental Citizenship.*

53 See, for example, John Dryzek, "Political and Ecological Communication," in *Debating the Earth*, ed. John Dryzek and David Schlosberg, 633–46 (Oxford: Oxford University Press, 1999).

54 George Lakoff, "Why It Matters How We Frame the Environment," *Environmental Communication* 4, no. 1 (2010): 70–81; Tom Crompton, *Weathercocks and Signposts: The Environment Movement at a Crossroads*, WWF report, 2008, http://assets.wwf.org.uk/downloads/weathercocks_report2.pdf.

55 For example, Ziad Abu-Saud, "The Dogma of Advertising and Consumerism," *Huffington Post*, January 25, 2013, http://www.huffingtonpost.co.uk/ziad-elhady/the-dogma-of-advertising-_b_2540390.html.

56 Tim Jackson, *Prosperity without Growth: Economics for a Finite Planet* (London: Earthscan, 2009), 153.

57 Richard Wilkinson and Kate Pickett, *The Spirit Level: Why Greater Equality Makes Societies Stronger* (New York: Bloomsbury Publishing, 2011), 230. Elegantly reversing the failed logic of trickle-down economics, which proposed economic growth as a substitute for equality, Wilkinson and Pickett observe that conversely, "greater equality makes growth much less necessary." Once recognition of the detrimental effects of inequality on human flourishing is added to the clear ecological impossibility of infinite economic growth, the pursuit of growth as a surrogate for equality looks even less attractive as a principle of social organization than it already did. The real task at hand is to reorganize over consuming societies such that everyone can flourish within a globally equitable ecological footprint.

58 Richard Layard, *Happiness: Lessons for a New Science* (London: Allen Lane, 2005); Barry Schwartz, *The Paradox of Choice: Why More Is Less* (New York: Harper Perennial, 2005); Jackson, *Prosperity without Growth*; Wilkinson and Pickett, *The Spirit Level*.

59 Wilkinson and Pickett, *The Spirit Level*, 252–63. More comprehensive systemic solutions may involve combining "degrowth" with some form of unconditional basic income—a significant challenge.

60 Mike Hannis, "After Development? In Defence of Sustainability," *Global Discourse* 7, no. 1 (2017): 28–38.

*Part V*

# EMBODIED CREATURE CONNECTIONS *to* OTHERS *and* PLACE

# TEN

# "Owning Up to Being an Animal"

## On the Ecological Virtues of Composure

DAVID W. JARDINE

## Preamble

MY SUGGESTION HERE, OF COMPOSURE, OF COMPOSITION, of writing as an ecological virtue, is not offered as an indiscriminate, universal aspiration. It is not necessary, only possible. The very declaration of universal aspirations is, in its own way, part of how we got into this eco-critical fix we're in.

My aspirations are not universal. Breath is always *someone's*. I own up. Every suggestion of ecological virtue is, at the same time, a wee plea for forgiveness. My capacity, like anyone's, is limited:

> When we find ourselves in dangerous situations in which there are abundant
> stimulants for the afflictions, we should cultivate the antidotes to them with
> a proportionate intensity, and we should stand up to them in a thousand

ways. It is said that the best practitioners use as the path the very object that gives rise to the afflictions. Average practitioners apply the antidotes and hold their ground. Practitioners of a more basic capacity must abandon such objects and retreat.[1]

I'm often unable to stand up in a thousand ways. Sometimes I must abandon and retreat and do what I then am able. When I do, I write.

In the end, "I compose this in order to condition my own mind,"[2] and then battle with trying to live up to things I can write about that are more ecologically virtuous than I can currently consistently live. Thus, too, Tsong-kha-pa's great reminder: "Practice those things that you can practice now. Do not use your own incapacity as a reason to repudiate what you cannot engage in."[3]

Writing, composition, composure—this is a small, specific medicine, but not for all ills or all occasions, not for all hands and hearts. And, after all, we've all gathered here in this book to compose ourselves and compose something about the composition of ecological virtues. We somehow all agree, however slightly and silently, that to think and rethink such matters, and to write and rewrite might itself be, in some small way, ecologically virtuous.

# I

The main title of this chapter comes from David Abram's beautiful book *Becoming Animal: An Earthly Cosmology*:

> Owning up to being an animal, a creature of earth. Tuning our animal senses to the sensible terrain: blending our skin with the rain-rippled surface of rivers, mingling our ears with the thunder and the thrumming of frogs, and our eyes with the molten sky. Feeling the polyrhythmic pulse of this place— this huge windswept body of water and stone. This vexed being in whose flesh we're entangled. Becoming earth. Becoming animal. Becoming, in this manner, fully human. ... Becoming a two-legged animal, entirely a part of the animate world whose life swells within and unfolds all around us.[4]

These words are so alluring. I have these green dreams. To be drawn into this blending. To deeply feel these pulses. To claim this entirety—the woozy, penumbral eco-greening of spiny nerve endings: being *of* the earth, embraced and embracing. And to claim thus the common strength of it, the comfort of this cuddle, its thrumming and mine. Yes.

I have these green dreams of rain ripple and mingle, of pulse and swell. Their pleasures are undeniable.

But what good are they? What is their virtue? Would that I could stay poised down in this embrace, forgetful of such questions, walking and breathing under this last foot of dark solstice snow, and nothing besides. I know full well, from painful and repeated experience, that this is a long, hard practice, gaining this poise. Some threads of ecological talk suggest that we let our strange humanity give way to the yield of this earthward practice, and that this is the path to ecological virtuousness.

Some suggest being animal.

This passage from David Abram doesn't quite say that. It suggests *owning up*. It also suggests this *in writing, beautiful* writing carefully composed.

This composure can be part, I suggest, of such owning up. Declarations, manifestos, urgent missives, and wailings of grief are needed in this owning up. But so, too, is the art of composure and composition: "Giraffes and tigers have splendid coats. We have splendid speech,"[5] providing we cleave with calm, composed care to our earthly composition.

## II

Even though it might be possible to, as goes the saying, make a virtue of necessity, something is not virtuous if it is inevitable. However dire our current circumstances might be, action in response is not necessary. In fact, the virtue of something virtuous is that it has no necessity to it. It must be chosen, "actualized." Owning up is needed only if it is possible to forget, to deny, to ignore, to lie, to withdraw and turn away, to repress, to be misled, or to misconstrue. As Abram's words indicate, our earthly, animate remains can fall from memory. Our earthly mingling can be left uncultivated, unnoticed, even disparaged as sin or contamination or goofy, unrealistic, liberal tree-hugging. Our skin's feel of its earthly fabric, left unblended and untended, can atrophy. And yet, all the while, our earthly being and that of our relations thrums along at its own pace, regardless of our regard of it, regardless of owning up or not, affected by our actions even if we deny those effects and lose our affection. This is why ecological virtue is not what we've done, what we do, or what we will do, but what we *ought* to do, *even if we don't.*

This is part of owning up, this particular thread of my animal earthliness. Distraction and affliction and forgetfulness are always nearby, as near as the rain thrum—nearer, often. If I am animal, a creature of the earth, then this, too, is part of that creatureliness. I can, I have, I continue to become busy, distracted, forgetful, fearful of thunder, linking the polyrhythmic pulse to night-startles of death, of suffering. And, because of the perpetuity of human suffering, it is easy to be drawn, in response, into marshals of "perpetual war"[6] against our earthly, animal, living circumstances. We can retract and begin to act as if our dens of retreat can become impervious to choke in the air and flint in the water, forgetting altogether how our unmeasured, panicky actions are themselves not just the *effects* of feeling threatened but are, in fact, part of the *cause* of that very choke and flint. Worse yet, the worse the air and water, the worse can be the threat-based retraction that cues of our large, prehensile brain to build larger and larger buttresses against the threat with little or no notice of how what we build can have a hand in causing the threat we then build against.

If we are creatures of the earth, this is part of our peculiarly human locale in the variegated array of animal fleshes. We lean toward the arising of a war against threat itself. It is perhaps no coincidence that our current ecological circumstances and the so-called War on Terror have co-arisen.

Owning up means, in part, owning up to how we have been perennially driven by this panic, this fear, how we've become repeatedly caught in a loop, where panic breeds panic, where threat breeds monstrous responses that then threaten us, as so many current ecological warnings are foretelling.

There is no sense trying to think about our circumstances, to think about what might constitute ecological virtue, from *inside* this self-perpetuating loop.

To own up to this, we need to compose ourselves.

## III

Owning up to being animal means owning up to the impermanence and perennial, repeated suffering of the world, human and non-human, and learning to not flinch, to not panic at this prospect, as if we could outrun it with our cleverness. As if outrunning is sane and virtuous and not simply exhausting and distracting. We cannot outrun our being animal with coal production, nor with ecological resistances to such production. Differently put, ecological resistance, as much as coal production, must own up to being an animal, to being a creature of earth, and therefore a creature bound to suffering and impermanence *no matter what we do or say or realize or fight for.* Realizing and responding to *this* with composure and grace is the great task of ecological awareness, what distinguishes it, and what makes David Abram's words far more burdensome and beautiful than a nice walk in the woods. Ecological resistance and action must think with terrible composure and not fall prey to flinching panic.

We need to compose ourselves in the face of this insight and learn to face it. I need to learn, again and again, of the earthly composition of breath and its coupling with the composition of those birds swooped at the feeder, with that young dog killed a few years back by the cougar and the sorrows that still arise, with this aging pinch of old skin on the back of my hands, and the brisk of a coming chinook as the sun lows, mooing milk-white near the horizon, solstice and the hoped-for coming return of light. To learn, again and again, of the frail winning of my own composure.

How is this earth-fabric composed, and how am I composed of it? Who speaks of it? What stories tell of it? To compose myself, I surround myself with composers who can read my life back to me from out beyond my own distraction and frailty. I read David Abram, for example, or Don Domanski:

> sunlight bright on pine boughs      saints asleep in the Great Bear
> the Great Bear asleep in the North Mountain      everything waiting....[7]

Or Michael Derby:

> Now little riverbed stones impress upon my bare feet the aggregate intelligence of form and fit, particular trees stand tall in my memory as pedagogically significant, the cheap yellow paint on my pencil peels and reveals flesh—what kind of mushrooms are these? From somewhere deep within the inquiry, beneath the words—how is it possible!—a world approaches.[8]

So, I walk and breathe (this, frankly, is tough enough by itself, to not become distracted in such practices), but I, like Abram, Domanski, and Derby, also

practise writing—composition—as a way to face the fix we're in, this imperilled fix of walking, of breathing, of seeing Great Ursula bristling in the minuses
of updraught air chills at night:

> Seeing the frailty of your life through seeing the breath is the meditation on
> the recollection of death. Just realizing this fact—that if the breath goes in
> but does not go out again, or goes out but does not come in again, your life
> is over—is enough to change the mind. It will startle you into being aware.[9]

But why, then, compose? Why struggle with the written word, here, now?
Why burden you as a reader with reading such words? Why isn't noticing
the ice crystals in intimate display overhead and feeling the animal shivers in
response enough?

Because, over the past thirty-five years, I've walked hallways of schools that
are breathless and have no repose. I've too often felt the pulse—of teachers,
of students, of my own—quicken in panic over threats of accountability, sur-
veillance-driven rushes, with affection lost for what is learned, for these living
fields of work we have inherited. Writing well of these matters can let our affection for these living fields grow. It can help make memory last.

Let me own up: it can, it *might*, outlive me.


# IV


In writing, we have a way to own up to part of our animality—we compose
about our composition, and some texts work as instructions on how and why
to remain under the bristle of stars, even when entering a school's enclosure:

> "Texts are instructions for [the] practice"[10] of precisely paying more intimate
> and proper attention to the resounding. Don't worry. Study, properly prac-
> ticed, will not ruin the *aesthesis* of ecological reveries, only their limited and
> limiting naiveties.[11]

Such well-composed works can thus help me remain composed. They remind
me that schools can be locales of teachers and students composing their lives
and learning of the composition of things, their incantational grammars:

The aim of such meditations is the cultivation of the intimacy and imme-
diacy of the experience of everyday life. Not only is "wherever you are...a
place of practice."[12] Tsong-kha-pa also insists [that]...the purpose and
object of study is precisely the deepening of practice itself. After all, "why
would you determine one thing by means of study and reflection, and then,
when you go to practice, practice something else?"[13]

[Composure, composition, is] meant, in the end, to make us more susceptible to the beautiful abundance of things as we walk around in the
world.[14]

To cultivate this work of insight into our composition, our composure, the
knife-edge fray of the world that would draw us into simply reacting to its
impingement must be held at bay. School, scholar: from the Latin *schola*, meaning, in part, "a holding back, a keeping clear."[15] It also, ironically, means "leisure."
Returning school to its Latinate roots of composure—composing myself in the
face of coming to know of the composition of things—has been, in my own
work, an ecological matter.

In the end, a weird obligation is placed upon the composure of scholarship,
to not waste this "leisure time" squandering one's breath while others' breaths
are robbed in thin air. Thus, the particular ecological composure of scholarship
itself requires its own tough, unflinching scrutiny. It is, in its own way, opulent,
luxurious, privileged, and it had better, therefore, earn its keep:

Why would I waste...such a good life? When I act as though it were insignificant, I am deceiving myself. What could be more foolish than this? Just
this once I am free from continuously trekking the many narrow cliff-paths
of leisure-less conditions, the miserable realms. If I waste this freedom and
return to those conditions, it would be similar to losing my mind.[16]

It is difficult to find ways to share what I can of the relief from suffering that
composure can bring, the ways it can clarify action, identify hidden conditions
and causes and antecedents that have blocked insight, reveal unspoken or forgotten or suppressed histories and aspirations. Writing. Study. These both rely
on and give rise to composure and thus, like any practices worthy of the name,
they require practice to become practised in.

# V

Owning up to our "being an animal" is thus, in part, *owning up to the weakness of that animality* and how it can distort our vision as much as clarify it, depending on how well we have learned to not be just drawn into its grieves and panics and startles and starts.

Too much ecological talk asks us to grieve and panic and startle and start. Like this: "alarms," "rapid deterioration," "egregious effects," "darkening shadows," "starkest warnings," "severely and irreversibly compromised," "unprecedented peril" (these are just some of the phrases used in the book proposal for this text). We can all add to this list with great (un)ease. Time is running out. Hurry. Point of no return. Poison. Degradation. Loss. Endangered. Extreme. Drought. Fires. Too late. Too late. Point of no return. Now. Please. Act. Stop. Help.

Every word becomes Capitalized. Then every letter.

Then bolded.

Then bigger in pitch and font and full of ever-multiplying exclamation points. And the email warnings arrive faster, ever-faster, ever-faster, ever-faster, every-single-one-now-marked URGENT!, their sheer and utter frequency— no, *acceleration*—outpacing the possibility of careful attention.

Without being able, somehow, to compose ourselves, we don't suffer simply disappearing ice and bees. We also suffer the environs of our own discomposure over such disappearance. Without composure, we're imperilled twice—first by our circumstances and second by the animal-spittle startle responses. Our circumstances impend suffering, and our fretful, heated, alarming language *about* those circumstances simply impend the impending of it:

> You must accept [suffering] when [it] arise[s] because (1) if you do not do this, in addition to the basic suffering, you have the suffering of worry that is produced by your own thoughts, and then the suffering becomes very difficult for you to bear; (2) if you accept the suffering, you let the basic suffering be and do not stop it, but you never have the suffering of worry that creates discontentment when you focus on the basic suffering; and (3) since you are using a method to bring even basic sufferings into the path, you greatly lessen your suffering, so you can bear it. Therefore, it is very crucial that you generate the patience that accepts suffering.[17]

Bringing our ecological suffering into the path of insight, composure, and clarity is how we own up to being an animal, to being *this sort of animal*. And, even though everything around us conspires to tell us that time is running out:

> *This can't be hurried.* Learn all you can about where you are, make common cause with that place, and then, resign yourself, become patient enough to work with it over a long time. And then, what you do is increase the possibility that you'll make a good example. And what we're looking for in this is good examples.[18]

I understand. Here I am, old enough to have the end clearly and constantly in sight, suggesting something terribly difficult to those who will likely suffer longer and more than I: This is the dreadful position that young people are in and I think of them, and I say that the situation you're in now is going to call for a lot of patience, and to be patient in an emergency is a terrible trial.[19]

Again, this terrible trial is part of the deeply human lot of owning up to being an animal. We must wrench ecological virtue from the grip of our own woes, our own suffering, and not fall prey to the modern sins of acquiescence, complacency, or denial that define far too much educational theory and practice, caught as it too often is in the hurry.

# VI

Here's a nightmare that haunts me too often: that both those protecting the earth and those despoiling it can be caught in analogous loops of discomposure, each pitted against the other, one monstrous prospect summoning up its equally monstrous opposite, meeting in kind over night terrors, each feeding the other, *needing* the other, unwittingly sustaining the other in order to sustain itself. There is no virtue to be had in maintaining ourselves in this particular dance of hubris and nemesis, and the long arguments over which is which. This is such an old story in human history, of one rising up to meet the other in kind, each convinced that they are sheer and utter opposites.

I'm reminded both of Friedrich Nietzsche's characterization of how human will deliberately seeks out that which resists it in order to increase its own feeling of agency and efficacy.[20]

I'm reminded, too, of Ivan Illich's terribly acute idea of "apocalyptic randiness: 'I have an even more horrible example to tell you! Let's imagine an even worse situation!' "[21]

Edward Said calls this "vocabulary of giantism and apocalypse."[22] Chet Bowers calls it a "Titanic mind-set."[23]

This tale—of the monster feeding on fear and growing in proportion to the (unwittingly) loving attention that the fearful give it—is as old as the hills.

Pausing over writing this, I'm asking myself, Am I in love with my panic? Does it make me feel alive and useful and "connected?"

So, what can I do to ensure that the monster won't get me, then? *Realize that there is no monster.* There is only the task, the choice, the possibilities, the work in front of me. "Save the Earth" makes it hard to save the earth beneath my feet. Just like "Literacy Skills" (too easily spoken in some schools with agonizing worry and aghast-ness) distracts from the real eco-pedagogical choices to be made, now, with these young children in this classroom, them with their joys and woes, and me with mine, gathering together to huddle over this story, this conversation, "this and this."[24] For pedagogical practice to be composed, it must not think globally and act locally, but must *think locally*, because it is always and only over the particularities of our lives that we live. The choice of what is best to do is small, intimate, here, nearby. Beware the monstrous shouts of Literacy Skills and Accountability and Surveillance and Time Is Running Out. Bloated capitalizations can easily overwhelm and make seem ridiculously meagre this small, intimately ecologically virtuous classroom event.

This is why, when speaking to teachers in schools, I often ask: "Who profits from our panic?" I've taken to repeatedly quoting from an interview with Kevin O'Leary, a former CEO of an online publishing company that makes various reading materials for use in schools with children: "I love the terror in a mother's heart when she sees her child fall behind in reading. I made a fortune from servicing that market."[25]

In order to maintain my composure in the fray of schools, I must understand with as much precision as possible that composure is surrounded by regimes bent on *deliberately* disrupting it and *profiting from that disruption*. And I must come to understand these regimes in enough detail to measure up to their arrival so I can decode them and unravel them when they loom up in monstrous display. This is the calm, predatory animal eye of scholarship. I compose myself in their presence, I name them, I cite, with great scholarly care,

I summon the strength and wisdom of my companions and ancestors to the best of my ability. I cultivate this ability. I practise. I compose, again and again, so that I and others might learn how to spot, how to evade, how to not fall prey to such deliberately induced terror.

## Notes

1   K. Pelden, *The Nectar of Manjushri's Speech* (Boston: Shambala Books, 2010), 253–4.

2   Tsong-kha-pa, *The Great Treatise on the Stages of the Path to Enlightenment (Lam rim chen mo)* (Ithaca, NY: Snow Lion Publications, 2000), vol. 1: 111.

3   Tsong-kha-pa, *The Great Treatise*, vol. 1: 49.

4   D. Abram, *Becoming Animal: An Earthly Cosmology* (New York: Vintage, 2011), 3.

5   J. Hillman, "Human Being as Animal Being: A Correspondence with John Stockwell," in *Animal Presences*, ed. J. Hillman (Putnam, CT: Spring Publications, 2008), 164.

6   D. Postel and S. Drury, "Noble Lies and Perpetual War: Leo Strauss, the Neo-Cons, and Iraq. Danny Postel interviews Shadia Drury," *Open Democracy*, October 15, 2003, https://www.opendemocracy.net/en/article_1542jsp/.

7   D. Domanski, "Madonna of the Diaphanous Life," in *Bite Down Little Whisper* (London, ON: Brick Books, 2013), 9.

8   M. Derby, *Towards a Critical Eco-Hermeneutic Approach to Education: Place, Being, Relation* (New York: Peter Lang, 2015), 2.

9   A. Chah, *Being Dharma: The Essence of the Buddha's Teachings* (Boston: Shambala Press, 2001), 44.

10  Tsong-kha-pa, *The Great Treatise on the Stages of the Path to Enlightenment (Lam rim chen mo)* (Ithaca, NY: Snow Lion Publications, 2004), vol. 2: 52.

11  D. Jardine, "Introduction: How to Love Black Snow," in *Towards a Critical Eco-Hermeneutic Approach to Education: Place, Being, Relation*, ed. M. Derby (New York: Peter Lang, 2015), xxii.

12  Tsong-kha-pa, *The Great Treatise*, vol. 2: 191.

13  Tsong-kha-pa, *The Great Treatise*, vol. 1: 52.

14  D. Jardine, "In Praise of Radiant Beings," in D. Jardine, *In Praise of Radiant Beings: A Retrospective Path Through Education, Buddhism and Ecology* (Charlotte, NC: Information Age Press, 2016), 304.

15  See "school (n.1)," *Online Etymological Dictionary*, https://www.etymonline.com/word/school (accessed February 23, 2020).

16  Tsong-kha-pa, *The Great Treatise*, vol. 1: 121–2.

17  Tsong-kha-pa, *The Great Treatise*, vol. 2: 172–3.

18  W. Berry with B. Moyers, "Writer and Farmer Wendell Berry on Hope, Direct Action, and the 'Resettling' of the American Countryside," *Yes Magazine*, October 11, 2013, http://www.yesmagazine.org/planet/mad-farmer-wendell-berry-gets-madder-in-defense-of-earth. Emphasis mine.

19  Berry with Moyers, "Writer and Farmer Wendell Berry."

20  F. Nietzsche, *The Will to Power* (New York: Random House, 1975), 346.

21  I. Illich with D. Cayley, *Ivan Illich in Conversation* (Toronto: House of Anansi, 1992), 127.

22  E. Said, "The Clash of Ignorance," *The Nation*, October 4, 2001, http://www.thenation.com/doc/20011022/said.

23  C. Bowers, *The False Promises Of Constructivist Theories of Learning: A Global and Ecological Critique* (New York: Peter Lang, 2005), 11.

24  B. Wallace, *The Stubborn Particulars of Grace* (Toronto: McClelland and Stewart, 1987).

25  *Dragon's Den*, series 6, episode 19, aired March 14, 2012, on the Canadian Broadcasting Company. Accessed online at www.cbc.ca/dragonsden/pitches/ukloo.

# ELEVEN

# Worthy of This Mountain

## Living a Life of Friction Against the Machine

DAVID CHANG

## Encounter

I N THE SUMMER OF 2016, MY WIFE AND I WENT ON A CRUISE TO Alaska. I was ambivalent about the trip, with its opulent parade of entertainment, all of which seemed bland replacements for the unmediated encounters that I had experienced on previous backpacking trips. Despite my misgivings, there were moments of awe: we marvelled at great columns of ice that split from the magnificent Hubbard Glaciers, the sound of the calving ice rumbling throughout the bay; pods of humpback whales following the ship's course, their barnacled backs slicing through black water, the spray of misty breaths puncturing the ocean surface.

The most salient encounter came when the ship sailed southward to Vancouver. As the late afternoon wore into evening, the sun boring through clouds that overlaid the stark waves, I looked out the port side and caught the sight of an imperial shape. It was Mount Fairweather, looming large above smoky clouds, standing in majestic resolution, basking in the fading light of day. Its grandeur drew me from my seat, and I ventured out onto the deck, braced against wild wind, and locked my gaze on the mountain's imposing stature. Its snowy robe glowed bright against the sun, a blaze of colour against the pewter dusk. The silvery twist of Fairweather Glacier rounded the foot of the mountain, pushing toward the gulf with mute force, bold under the clouds' shadows. It was an engrossing scene: the audacity of rock, the strength of ice, the wizened cap, the softening sky, the roiling waters. I stood in rapture, at once breathless and bursting with gratitude. The scene stole upon me without the slightest pronouncement—there I was, rapt in the utmost marvel, as if witnessing my own Sinai. Fairweather was a gift, and it seemed a miracle that I should have been there to witness its splendour.

But the wind grew bitter, and the ocean chill started to descend. I shuffled toward the door. Having once relinquished the sight of a rare eclipse, Annie Dillard wrote: "one turns at last even from glory itself with a sigh of relief."[1] As we left Mount Fairweather in the distant shore, I did not feel relief, but rather a stirring charge. I had witnessed a splendid manifestation of the earth's power in elegant form. The privilege of that encounter was deeply humbling. *Who am I to behold this majesty? Why should I be here to experience this numinous sight?* It baffled me that I should be so fortunate to behold this marvel. And as with most responses to unmerited grace, I considered how I might respond to this gift. At the very least, I should muster some calibre of being that is becoming of the mountain's dignity. *How will I live a life that is worthy of this mountain?*

## Exposition

Many notable ecological theorists have related "moments of conversion," when a salient experience broke through the skin of inveterate views and initiated a new ecological consciousness. Aldo Leopold found his definitive moment as he watched the dying fire in a wolf's eyes.[2] Paul Watson committed his life to the protection of oceans when he gazed into the eye of a sperm whale as it rolled onto its side, a harpoon lodged deep in its flesh.[3] My own encounter with Fairweather was not, as with Leopold and Watson, the pivotal moment that marked the end of one life and the beginning of another, but rather a lucid recognition of long-held values culminating in a moment of illumination. Over the years, my deepening appreciation of natural beauty coincided with a growing concern over the plight of the earth. As my own ecological awareness expanded, so has my sensitivity to moments of gratuitous delight. Mount Fairweather was the vision of the sublime that reinvigorates an ethical purpose. The aspiration to live a life worthy of the mountain stems from a desire to equal in virtue and excellence the mountain's inalienable beauty and dignity, and to commit myself to the cause of planetary vitality, of which the mountain is one manifestation.

Yet if the mountain inspires any injunctions on the conduct of a human life, it is not immediately clear what ethical directions I was to draw from the mountain. What does "living a life worthy of the mountain" look like? To start, I can read the mountain symbolically as an emanation of the earth's splendour, a touchstone of beauty that unfurls from the abundance of the planet. In this sense, Mount Fairweather instantiates a larger, undivided *gestalt*[4]—the earth itself—in all its vibrant diversity. To live a life worthy of the mountain is thus coterminous with a determination to live in harmony with the planet. On the other hand, if I take Mount Fairweather as an entity in itself, without extrapolating any global significance, I must recognize its unique and imminent beauty. Aldo Leopold once wrote: "A thing is good when it contributes to the stability, integrity and *beauty* of the biotic community."[5] Because my experience of Fairweather was undeniably aesthetic in nature, my ethical commitment would therefore have me preserve the beauty of the mountain as I experienced it that evening. For this reason, the injunction surfaces in a call to conservation, the protection of the mountain's snow cap, its surrounding glaciers, and its neighbouring watersheds.[6]

Both these ethical directions press for challenges to my urban life. I live far away from the mountain, but my participation in a vast, modern civilization predicated on unbridled growth, rapacious consumption, and unremitting exploitation contributes to the compounding threats that undermine this mountain. The gasoline that combusts inside the engine also heats the atmosphere that now thaws Fairweather's glaciers; industrial effluent sent adrift in the air now settle on the pristine banks of the mountain's cap. Amidst the drone of congestion and construction, the human city hums with blithe indifference. My complicity in an insatiate industrial system means that I have a hand in ecological harm, even if such harm is inflicted through no deliberate malice.

If I am to live a life that responds to this mountain, I must, in the very least, interrogate the human systems that threaten the mountain. However, to rile against the city and the system it represents is to inveigh against my own urban conditioning. How am I to live out the inspiration of the mountain while living in the city?

## The System Within and Without

A friend once showed me an amusing video of a toddler leafing through a magazine.[7] The baby had grown accustomed to tapping and swiping the glossy screen of her iPad. With the magazine spread before her, she taps her fingers on the page, expecting an instantaneous response. When the page remains inert, the child presses harder, spreading her fingers across the paper. Her hapless attempt at summoning a response made for delightful hilarity; but the scene was both endearing and revealing. To the extent that humans have created technology to aid and "enhance" human life, technology has in turn "given birth" to a certain kind of human by altering our perceptions, regulating our actions, and shaping our inclinations. That a child should expect paper to behave like pixels intimates the ways in which technology encroaches upon biology. The child transfers a mode of interface to the physical world—the iPad has instilled a code of operation and has replicated itself in the child's interactive modality. We see in her attempt at coaxing a response from the page the actions of a technological creature whose behaviour re-enacts the functional parameters of a machine. The machine has become a source of mimesis, a frame from which ontogeny proceeds.

This is not to say, of course, that the child will not eventually learn the difference between paper and tablet. However, the application of technological know-how to matters outside computer technology demonstrates a salient feature of the tool-making animal. Our machines help us master the world; but the more we use machines, the more we expect the world to conform to the logic of the machine. In urban North America, few areas of modern life are left untouched by mechanization and computerization. The effectiveness of machines produces an expectation of efficiency that seeps into areas of human life beyond that of material production. Symmetry and linearity, the engineering principles of mechanization and computation, are inscribed into the design of cities, the construction of shelters, and the arrangement of institutional order. Such patterns not only schematize social relations, they entrain our bodies by determining our material surroundings and circumscribing the scope of sensual contact. Rigid lines and right angles calibrate our visual perception, inculcating an underlying aesthetic sensibility; flat surfaces and smooth textures engender a baseline of sensual ease. Reliance on motorized vehicles raises expectations of a range of travel while diminishing our physical ability to walk short distances. The prevalence of light, emitted through bulbs and screens, effectively prolongs the day and alters hours of activity, the duration of mental stimulation and patterns of sleep. Machines help us live in the world, but we now live inside the world of our machines. Each of our technological inventions have, in ways both subtle and significant, asserted the terms of our tenure as a species on the planet, and we have, in turn, become the subjects of our technological apparatus.

If we writ large the machine as a metaphor of modern industrial economy, other parallels arise. The machine is a system of integrated parts whose functions coalesce to serve a purpose. Each constituent part performs a distinct operation—but the coordination of disparate operations constitute the singular function of the machine. Similarly, a modern economy operates on divisions of labour, the collocation of roles that comprise cycles of production and consumption. A worker in a modern economy contributes labour in exchange for wages that buy a range of goods and services, all rendered at the hands of other producers. Elaborate systems of extraction supply the materials from which consumer goods are made. By participating in the economy, we are effectively agents of production and consumption; while we may have some control over our own livelihood, we do not effectively wield the same agency over the occupations of others. A cog in a machine does not concern itself with other gears—in performing its own task, the cog keeps to its design, just as other components work strictly to their assigned tasks. The system works because people do their part. To be a part of the machine is to recognize that we are all complicit in keeping the machine running. No one person is ultimately in charge; no agency is absolute.

Recognizing the current form of modern industrial economy as a deleterious force on the vitality of the planet, resistance against the machine[8] becomes imperative. In his treatise on civil disobedience, Henry David Thoreau enjoined: "let your life be the counter-friction to stop the machine";[9]

moreover, he advised that we not "lend [ourselves] to the wrong which [we] condemn."[10] The injunction presumes, however, that the operations of the machine are entirely external to us, that we stand apart from the system we resist. What if the system has already occupied our inclinations, dictated our habits of thought, and prescribed our range of actions? In this case, our efforts to resist the system may reinforce its very strength and tenure, despite our best intentions.[11] The "machine" that Thoreau railed against was a government that sanctioned slavery, which he saw as the cardinal malfeasance of the American state; for us, the root source of the ecological crisis might stem from an industrial capitalist order, a culture predicated on materialism and consumption. Active resistance might include lobbying for changes in legislation to limit the power of corporations, the strengthening of environmental regulation, opposition to pipeline expansion. While these efforts may stem the egregious incursions of an already expansive system, we should also revitalize and reform our own cultural landscape, which involves the reinvention of traditions and the transformation of our own consciousness. In short, challenging the modern industrial capitalist machine requires work both inside and out.

## Habitus and Social Reproduction

If we examine the logic of the system in our own lives, we might see the development of abiding dispositions as a salient outcome of social reproduction. Pierre Bourdieu uses the term "habitus" to denote "a system of durable dispositions" that function as "principles of the generation and structuring of practices and representations which can be objectively 'regulated'…without in any way being the product of obedience to rules…without a conscious aiming at ends."[12] A system inculcates without explicit intention or conscious call to order—the regulation of practices flow from the function of institutions and the ongoing inertia of cultural life. The dispositions that make up habitus result from a person's life experiences within a social context; they are "durably inculcated by the possibilities and impossibilities, freedoms and necessities, opportunities and prohibitions inscribed in the objective conditions."[13]

The aggregation of dispositions fashions and regulates practices in a society in which a given habitus enjoys purchase. So the habitus is a product of social conditions, which itself produces practices that extend the social order. Bourdieu argues that habitus's enduring power lies in its invisibility:

> It is therefore not sufficient to say that the rule determines practice when there is more to be gained by obeying it than by disobeying it. The rule's last trick is to cause it to be forgotten that agents have an interest in obeying the rule, or more precisely, in being in a regular situation.[14]

In a manner of speaking, a habitus is the programing that underlies all social repertoires and patterns of behaviour, the *modus operandi* behind the *opus operandum.*[15]

Ingrained dispositions correspond to objective demands and partly constitute the capacity for social adjustment. Habitus enables one to function well in a given social reality; one's inclinations meet facets in an external field that produces those inclinations. Like a fish that does not feel the weight of water, a functioning habitus is not apparent to the agent: his dispositions are subsumed into *common sense*—which accepts reality as self-evident and beyond dispute.[16] Because "every established order tends to produce…the naturalization of its own arbitrariness,"[17] an obscure novelty can be met with consternation by the existing order, offensive as it is to ruling sensibilities.

Habitus is not simply a pattern of mental thoughts and attitudes, but a "meaning-made body."[18] The norms established by pervasive practices transfers onto the *soma*, such that physical comportment itself becomes an emanation of social meaning: "what is 'learned by the body' is not something that one has, like knowledge that can be brandished, but something that one is."[19] Habitus is inherent in somatic experience, and often lies beyond conscious discernment. The social world makes a nest of, and insinuates itself through, the body: "the habitus as the feel for the game is the social game embodied and turned into second nature."[20] The body is therefore the site of *hexis* (a state or habit), which is both the signifier of meaning and the meaning itself. Postures, gestures, patterns of gaze, and inflections of voice mark one's habitus; variations in preferences for pleasure and tolerance for discomfort belie the dispositions that are the products of society. Social divisions are also demarcated—class, gender, and age reveal themselves in the bodies of social agents through mannerisms and physical habits. The habitus of the body fashions a range of preferences and tolerances that circumscribe somatic mien. The immanence of habitus within the body *hexis* means that a challenge to habitus is likely experienced as an affront to the body itself.

The reproduction of dispositions promotes degrees of social cohesion through acculturation. Thus, the removal of a subject from her environs does not neutralize the effects of her habitus—she will continue to be compelled by the inertia of internalized dispositions. In so far as I live in society, society also lives in me.

## Hyper-Pessimistic Activism

In Pierre Bourdieu's theory of habitus, we see a means through which a social system is internalized and embodied in the social agent; friction against the machine from this perspective entails the reform of one's own inclinations and norms. On the other hand, ruling operations of the society form severe constraints on self-reformation. Techniques of cohesion, conformity, and discipline are institutionalized and administered throughout society, as are the forces that rationalize mass production and consumption, assigning occupations that determine the purpose of a human life. The nodes of coercive power have evolved with time, but the effects of disciplinary regimes have persisted despite many permutations. The confluence of tacit norms, institutional structure, cultural practices, and written law form a pattern of collective life that enforces a sociality that limits autonomy and integrity.

In *Discipline and Punish*, Michel Foucault presents a genealogy of discipline as a technique of social control.[21] He links the methods of the carceral state to the public executions in the French Republic, the practices of pedagogical stricture to the methods of control deployed in prisons. Foucault begins by visiting the grisly spectacle of the public execution, an event that marked French life in the seventeenth century. The authorities devised macabre methods of execution, most of which sought to inflict maximum suffering by dismembering and mutilating a prisoner's body. More remarkable, however, is the extent to which the body continues to be abused after the condemned has died. Justice pursues the body beyond death, Foucault argues, because an execution not only punishes the condemned but imposes discipline on the populace. As a public spectacle, the execution displays the Crown's coercive power: the mutilation of a dead body serves no other purpose other than to demonstrate the state's unbridled might. The public's riveted gaze becomes the channel through which power asserts its course. Foucault further argues that the gaze, as a channel of power, transmutes into systems of surveillance in prisons, where disciplinary effect is sustained without the overt application of coercive violence. In schools, the "supervision" of students recapitulates the surveillance techniques that mark the carceral state. Pedagogical strictures around pose and posture ("sit up straight," "eyes on me") re-enact impositions against the *soma* the way justice lords over a body in its exercise of power. Requirements on physical comportment is pedagogy's way of demanding deference to the establishment; it "extorts the essential by demanding the insignificant,"[22] thus enforcing a manner of submission among the student populace. According to Foucault's analysis, the force of blunt, coercive power has not disappeared; rather, power has dispersed and replicated itself through multifarious nodes of influence, propagated in new mediums of social relation.

Foucault's work is sobering for progressives who believe that tenacious activism and enlightened education undoubtedly bring about positive social change. In Foucault's view, the instruments of oppression continue to wield influence in their new forms. Some may find this view bleak and disheartening. Indeed, Foucault himself once characterized his philosophical corpus as a practice of persistent skepticism in response to the manifold technologies of society, an attempt to dismantle the suppositions that hold sway over us:

> my point is not that everything is bad, but that everything is dangerous, which is not exactly the same as bad. If everything is dangerous, then we always have something to do. So my position leads not to apathy but to a hyper- and pessimistic activism.[23]

The deconstruction of dominant modes of life pulls the rug from underneath us, as it were, and deprives us of our unqualified trust in given norms. A *hyper-pessimistic activism* is the condition of a freedom that resists captivity under the administration of culture and society. Its unrelenting questioning makes the very prospect of progress uncertain, because Foucault tears the subject from itself, such that "it is no longer the subject as such, or that it is completely 'other' than itself so that it may arrive at its annihilation, its dissociation."[24] In other words, we see ourselves entirely involved in egregious systems, and the systems active in us. The more we see this clearly, the more we recognize "the impossibility of living."[25] Hyper-pessimistic activism is therefore a lifelong engagement in struggle, a form of unremitting vigilance against the infiltration of deleterious power, injustices that subvert agency and integrity. "Pessimism," in this case, does not denote an inclination toward gloom, but rather defiance against facile innocence and unqualified optimism—it is an abiding sobriety that refuses to adjust to pervasive oppression and exploitation.

If one strives to live a life worthy of the mountain by challenging the systems that threaten its beauty, one must reform one's internal habitus, a form of conditioning under the aegis of modern society that prevents us from embracing a life of radical simplicity and deep respect for the earth, while also countering the external channels that reinforce a given way of life. What constitutes virtue, in this case, is the commitment to *becoming* a certain person, even if one's existing dispositions hamper that very commitment. At the same time, one recognizes that a person's very self is composed by an overarching social system, absent the reform of which no act of self-reformation counts for much. Friction against the machine, then, simultaneously entails the work of changing both the self within and the system without.

## Virtue in Practice

In discussions about virtue, the distinction between the system *within* and *without* serves as a useful descriptor of different domains of human life; in practice, they form two aspects of a single commitment. If we have internalized a given social norm, then the attempt to re-school our inclinations will come at the cost of our own comfort as well as the risk of social stigma. Consider a middle-class North American who adopts veganism due to ethical concern for the ecological consequences wrought by industrial meat production. Her ethical commitment may require that she deny her cravings; further, her ethical stance may put her at odds with cultural milieus in which meat is consumed without scruple. Veganism, therefore, will challenge her own habitus, as well as her position within a social web held by shared practices and rituals. The same might be said of a Canadian who refuses to own a car in order to extricate himself from the petroleum industry. If he lives in a place without an efficient transit system, his ethical commitment has implications for his comfort and convenience, not to mention significant disruptions to his life. In an economic system that relies on the flow of fossil fuels, a stringent refusal to burn gasoline may also have dire consequences for one's career possibilities. Thus, there is a personal price to pay for the cultivation of ecological virtue.

If we consider the price of virtue, we soon find that the friction we apply against the machine also grinds us down along with the apparatus we aim to stall. The mantle of counter-establishment activism is heavy, and often exacts a hefty psycho-spiritual toll. Efforts toward ethical integrity, the willingness to forego benefits afforded by an unsustainable way of life, can sharpen our awareness of complacency in others. The more we demand of ourselves and society, the more susceptible we become to frustration and indignation, or what Eamonn Callan calls "moral distress": "a cluster of emotions that may attend our response to words or actions of others or our own that we see as morally repellent."[26] Such distress is inevitable to the process of moral development, for "a man [sic] who is supremely compassionate cannot view with impassivity the many blameworthy ways in which humans fail to be compassionate nor can he regard his own failures in that light."[27] In this sense, probity risks the rupture of connection and community; the fastidiousness with which one pursues personal virtue also becomes a severe criterion against which the actions of others are judged.

My own wrangles with moral distress stem from my decision not to have children. I have always adored children; this appreciation in part led me to become a teacher, and as a young man I had always imagined myself having children of my own. However, in my travels, I saw much appalling poverty throughout the world; many children staving off hunger, struggling to eke out a living. I saw vast landscapes despoiled by human masses, rivers blackened by effluent, lakes covered in detritus. Much of this malaise stems from a warped economic system, whereby rich nations export their environmental damage to poor nations in exchange for cheap goods. There is no way to separate my standard of living from what I witnessed abroad. I realized that I could not in good conscience add further impetus to an unsustainable way of life by contributing to a society (Canadian) that demands so much of the earth's resources. There is enough work to be done simply addressing existing inequality, lack of access to education and health care among those *already here*, much less investing energy into raising my own progeny.

Many have asked me about *when* I planned to have children—a question that assumes by default that I *would* have children. People are surprised that I consciously decided against having children; and when they ask me for reasons, the conversations can quickly become uncomfortable. I do not always wax theoretical on the socio/ecological implications of population growth; however, I often feel that my personal choice upsets an underlying cultural order. These conversations instantiate moments of friction that strike against a prevailing norm.

Many of my friends are now having children of their own. Announcements of pregnancy are usually met with elation and celebration, and I am among those who offer my congratulations. At the same time, I find myself struggling with conflicting thoughts. *Do they know the state of the ecological crisis and the role that human beings play in the problem? Do they know that their children will have to live with nuclear waste that we have yet to deal with? Have they considered how climate change is likely to further destabilize societies already stricken with inequality and fragile political institutions?* Some will consider me paranoid and needlessly anxious; but to me these are matters that require consideration— part of the responsibility of being a parent.[28] Yet I cannot presume to judge others based on my ethical views, which are undoubtedly coloured by personal dispositions and experiences; a *moralistic* attitude is invidious and corrosive to community. Therefore, I do not explicitly question my friends' decision to have children. There is an inexorable part of me that adores children, but the joy I share with pregnant friends is never without concern. While my congratulations are offered with sincerity, they always mask a certain foreboding regarding our future plight. Thus, I harbour an internal tension as a result of having staked an ethical position, and my interactions with others are inevitably coloured by these convictions.

This is not to say, on the other hand, that I believe humans are a planetary plague, that as a rule the earth would be better off if there were fewer of us. Karen Litfin has cautioned against the "misanthropic temptation,"[29] which corrodes the foundations of community and the collaborative efforts required to move toward a healthier relationship with the planet. I do not forward a case against having children as a way to hone ecological virtue (thereby implying that those who have children are not virtuous[30]). Rather, I present my experience as an example of how working against the machine induces internal tensions. If I derive an ethical position by examining the implications of a collective lifestyle, and my commitment is motivated by a concern for the greater good, I will find it difficult to endorse the actions of others who appear to contravene my view of what serves that good. Magnanimity is easy in matters of subjective preference (I prefer the Rolling Stones; you favour the Beatles—to each his own). Matters of society are often thorny because they call on us to share life with others, whose views differ from our own, to live together in a world in which one's actions affect all.

To live in friction against the machine, then, requires that we address the disjuncture between a commitment to our own ethical values and the *perceived* shortcomings of others. To this end, a reminder from Aristotle proves instructive:

> Anything that we have to learn to do we learn by the actual doing of it: people become builders by building and instrumentalists by playing instruments. Similarly we become just by the performing of just acts, temperate by performing temperate ones, brave by performing brave ones.[31]

I read in Aristotle a particular emphasis on "practice"; it is the *doing* that instils excellence. Expert knowledge of music theory fails to coax a sound without the pull of bow over string. Similarly, virtue is abstract until it is reified in action, substantiated through practice, sanctioned through repetition. Taking Aristotle's point further, and with attention to *orthopraxis* (right practice) and *orthodoxy* (right belief),[32] I submit that the formation of virtuous character lies not so much in the insistence on discrete beliefs but in a commitment to beauty and excellence through action. A hyper-pessimistic activism that aims to disrupt the machine thus draws its vitality from consistent practice: a vegan cultivates compassion by making compassionate choices every meal; a teacher nurtures kindness by extending kindness to students.

Although the apposition of *orthopraxis* and *orthodoxy* may clarify contrasting inflections of virtue, we may find the distinction negligible in everyday life. After all, the consistent practise of good and beautiful activities inevitably insinuates to our minds a set of views about the good and the beautiful. The more I align my life with my ethical values, the more I become invested in the supposed merits of those values, which in turn galvanizes notions of how I ought to live. An adherence to *orthopraxis*, in other words, does not forestall the encroachment of *orthodoxy*.

The difficulty lies not only with the mutual influence of practice and belief: the status of virtue itself is called to question if we consciously recognize our efforts as virtuous. One can argue that a virtue that deems itself virtuous is no virtue at all, just as a person who prides herself on her humility has already fallen to conceit. So we come to a distinct conundrum that threatens to grind us down along with the machinery we want to disrupt. We must deliberately practise virtue without conscious awareness of our practices as ostensibly virtuous. We must actualize in *practice* what we see as good, excellent, and beautiful, without deriving a suite of *beliefs* that can ossify into dogma.

How do we navigate this delicate terrain? I suggest that the cultivation of virtue benefits from an attitude that strengthens the commitment to virtue while disabusing ourselves of any notion that we are virtuous. Personally, I derive guidance and inspiration from Buddhist practice, which systematically questions the perceived substance of an ego, virtuous or otherwise. The following instruction from Shunryu Suzuki regarding Zen meditation is germane to the practice of virtue:

> as long as you think, "I am doing this," or "I have to do this," or "I must attain something special," you are actually not doing anything. When you give up, when you no longer want something, or when you do not try to do anything special, then you do something.[33]

Suzuki's explication attempts to assuage the anxious fixations that muddle our efforts and return us to the purity of practice. He enjoins: "we must make some effort, but we must forget ourselves in the effort we make."[34] Thus, our labour can be seen as *nothing special*, nothing particularly notable. To soften one's grip on the parochial *ego*, however virtuous and wise, frees us from the tensions induced by self and other, virtue and vice, ethical and unethical. Although we have staked a position in working against the machine, we can do so without vexation. It is possible to pick up a serious task with the lightheartedness of a child, to shrug off our consternation with cool equanimity, to invoke humour in defiance of grim odds. Suzuki's words articulate insight in the work of shaping the self, one that can be likened to the finesse of an artist: we can bend and carve our lives into the very shape of virtue, but the finishing varnish dissolves the trace of virtue altogether.

I utilize two practices that help me to forget myself, that I am doing nothing special. When I catch myself ensnared by anxiety over the state of the world, or feel myself tighten up in the effort of "doing something ethical," I practise a body scan—I take deep breaths to collect my awareness and return my attention to the sensations of my body. I start from the top of my head and scan slowly, noticing bits of tension that I hold in my muscles. Usually, I observe tension around my neck and shoulders. With the fine comb of awareness, I sense that even the muscles around the face are often taut. As I detect tension, I relax each muscle, moving my attention to my chest, arms, abdomen, legs and feet. By practising this combination of attention and relaxation, I sense how mental stress is held in the body; as I soften my muscles, the mental struggles dissipate. In this way, I can apply myself to my work without vexation.

The other practice is to seek guidance from the earth itself. When I am assailed by tension, I turn toward some element of the natural world—the sky, the trees—and ask: *What should I do with this angst?* The question is not usually an explicit one—it can come in the form of an unarticulated intention. In essence, I lay my troubles under the sky and take cues from the earth's inherent wisdom. So what do the sky and the trees say? Nothing! They always respond to my plea with exactly what they already are. The trees keep swaying in the breeze, the sky remains silent in its spaciousness. Yet this is a sublime answer to my trouble. I can sense that by the sky remaining the sky, the trees the trees, they tell me to stop pursuing solutions and resolutions. I see my struggle as an artifice that obscures a fundamental, inviolable perfection that pervades the universe. Everything is sufficient; everything is complete. Everything is workable. So in listening to the earth itself, I can drop my struggles and feel supported by the world. The knots disappear. Once I tap into this boundless energy, fatigue fades. The work carries on.

## Return to the Mountain

Having toured the possible hazards of resisting the machine, I may have conjured too grim a picture of ecological work and the demands of living in the Anthropocene. In actual fact, a life dedicated to preserving and enhancing ecological vibrancy brings exuberant rewards. A person moved by the splendour of the earth will find solace in sunshine, in the lapping foam of the ocean's waves, the lilting cadence of the warbler's song. New affinities and associations emerge as one resolves to live life according to ecological values. Naturalist organizations, local garden collectives, community-supported agriculture, cycling coalitions, tool libraries, ecological restoration associations, and animal rescue shelters are a few examples where like-minded people are countering the industrial/consumer impulses by developing local culture and communities of compassion. Going further, one can experiment with ways to "opt out" of the establishment by going on media fasts (turning off the TV, computer, cell phones) in an effort to reclaim one's mental space; we can try solitary wilderness retreats to nourish the capacity for silence and awareness, which are under perpetual assault by the modern establishment. Few antidotes to cynicism and apathy are as effective as collaboration with like-minded others, working on realistic goals that reap tangible outcomes.

Since much of the current ecological crisis traces back to those who enjoy a middle-upper-class lifestyle, the onus is on the developed nations, especially those in the West, to curb consumption and to transform the cultural-economic landscape. This requires significant alterations to what we deem as the material requirements of a good life. We will have to dismantle our sense of entitlement as we shift away from heavy industry to local production on a smaller scale.[35] We will have to relinquish material abundance as a symbol of personal achievement and worth and embrace simplicity as personal freedom. This endeavour will challenge our ingrained habitus while demanding tenacious struggle against social structures. Perils await, as do indescribable rewards. But the practice of countering the machine can itself be a source of fulfillment—for although we shall not always see the fruit of our labour, we can imagine in the heat of our grunt and sweat the royal shape of Mount Fairweather. We can agree with Camus that all is well, that the struggle itself is enough to fill our hearts. Like Sisyphus, we can imagine ourselves happy.[36]

## Notes

1   Annie Dillard, *Teaching a Stone to Talk*, rev. ed. (New York: Harper Perennial, 2013).
2   Aldo Leopold, *Sand County Almanac* (New York: Ballantine Books, 1986).
3   *Eco-Pirate: The Story of Paul Watson*, directed by Trish Dolman (Vancouver: Screen Siren Pictures and Optic Nerve Films, 2011).
4   Here, I draw from Arne Naess's reading of Spinoza's metaphysics. *Gestalt* is the conception of an abstract, united whole, a construction of the rational mind (*ens rationis*) derived from the experience of the concrete and the specific. The sight of a mountain enters into Mountain as a mental category; likewise, the experience of beauty can be encompassed by Beauty, a schema under which all aesthetic encounters are subsumed.

5   Leopold, *Sand County Almanac.*
6   A critic might deem beauty a flimsy guide for ethical action. The notion of beauty is inseparable from subjectivity; thus, pinning ethics to the preservation of beauty may leave too much to subjective judgment. Further, some theorists have argued that wilderness conservation tends to favour grandiose landscapes that appeal to human aesthetic taste (for example, Yosemite and Banff) and overlooks less striking places like marshes and oceans (see John Rodman, "Four Forms of Ecological Consciousness Reconsidered," in *The Deep Ecology Movement*, ed. Alan Drengson and Yuichi Inoue, 242–56 [Berkeley, CA: North Atlantic, 1995]). In response to these critiques, I return to Leopold's triad: a thing is good when it tends to preserve the stability, integrity, and beauty of the biotic community. Beauty in itself does not serve as the sole guide for ethical conduct, but the consideration of how an action sustains the stability and integrity of a place adds further direction for ecological ethics.
7   UserExperiencesWorks, "A Magazine Is an IPad That Does Not Work.M4v," https://www.youtube.com/watch?v=axv-yaFMQNk (accessed August 18, 2017).
8   The machine can be thought a technological/capitalist/industrial one as well as an anthropological one. Giorgio Agamben, in *The Open: Man and Animal* (Stanford, CA: Stanford University Press, 2003), deems the "anthropological machine" a vast complex based on the presumption of an ontological distinction between humans and animals. The human is constructed precisely from that which animals are not. The anthropological machine is the formalization and systematization of anthropocentrism across many cultures.
9   Henry D. Thoreau, *Walden and Other Writings* (New York: Random House, 2000).
10  Thoreau, *Walden and Other Writings*, 677.
11  For example, the push for technological "solutions" to anthropogenic climate change has led to a growing exuberance around electric cars, which reinforces the capitalist regime that underwrites their production, and stokes the myth of the revolutionary genius (e.g., Elon Musk) as the instigator of rapid social transformation. Both are forms of unqualified optimism that tend to assign the task of transformation to technology and industry while absolving the public of hard activism, personal responsibility, and drastic alterations to lifestyle.
12  Pierre Bourdieu, *Outline of a Theory of Practice*, trans. Richard Nice (Cambridge: Cambridge University Press, 1977), 72.
13  Pierre Bourdieu, *The Logic of Practice* (Cambridge: Polity, 1990), 54.
14  Bourdieu, *Outline of a Theory of Practice*, 22.
15  Bourdieu, *The Logic of Practice.*
16  Pierre Bourdieu and Loïc J.D. Wacquant, *An Invitation to Reflexive Sociology* (Chicago: University of Chicago Press, 1992).
17  Bourdieu, *Outline of a Theory of Practice*, 164.

18  Bourdieu, *The Logic of Practice*, 43.

19  Bourdieu, *The Logic of Practice*, 73.

20  Pierre Bourdieu, *In Other Words: Essays towards a Reflexive Sociology* (Cambridge: Polity, 1990), 63.

21  Michel Foucault, *Discipline & Punish: The Birth of the Prison* (New York: Vintage, 1995).

22  Bourdieu, *The Logic of Practice*, 69.

23  Michel Foucault, *The Foucault Reader*, ed. Paul Rabinow (New York: Pantheon, 1984).

24  Michel Foucault, *Remarks on Marx: Conversations with Duccio Trombadori* (New York: Semiotext(e), 1991), 31.

25  Michel Foucault, *Remarks on Marx*, 31.

26  Eamonn Callan, "Virtue, Dialogue, and the Common School," *American Journal of Education* 104, no. 1 (November 1995): 1–33, https://doi.org/10.1086/444114.6.

27  Callan, "Virtue, Dialogue, and the Common School," 8.

28  What about those who have considered the collective implications of child-bearing but have arrived at conclusions different from mine? Although they have fully assumed the responsibility of careful consideration, their conclusion may be at odds with mine. While I can laud their sense of responsibility, the disagreement over the ethical implications of child-bearing can remain a source of tension and difficulty.

29  Karen Litfin, *Ecovillages: Lessons for Sustainable Community* (Malden, MA: Polity, 2013).

30  A friend once said to me, rather facetiously: "Dave, you should absolutely have children. Your kid will be the next David Suzuki!" I blush at the suggestion. To agree with my friend is to presume too much about my parenting skills, as well as to hold a deterministic view of child-rearing, requiring me to impose onerous expectations of what a child ought to be, the life she ought to live.

31  Aristotle, *Nicomachean Ethics*, 2nd ed. (Indianapolis: Hackett Publishing, 1999), 1103b1, 92.

32  Aristotle, of course, does not dismiss outright the importance of wholesome views. In the final book on the Nichomachean Ethics, he returns to the matter of happiness, positing that the activities that accord with the noble and divine, the insights that spring from the best of our nature, are considered the highest happiness (eudaimonia). Aristotle believes that contemplation (the practise of wisdom via the intellect) is not a means to a higher good, but is itself the highest good, since the wisdom needs no further justification beyond itself.

33  Shunryu Suzuki, Richard Baker, and Huston Smith, *Zen Mind, Beginner's Mind: Informal Talks on Zen Meditation and Practice*, ed. Trudy Dixon, rev. ed. (New York: Weatherhill, 2000), 47.

34 Suzuki, Baker, and Smith, *Zen Mind, Beginner's Mind*, 37.

35 E.F. Schumacher, *Small Is Beautiful: Economics as If People Mattered* (New York: Harper Perennial, 1989).

36 Albert Camus, *The Myth of Sisyphus: And Other Essays* (New York: Vintage, 1991).

# TWELVE

# Stories of Love and Loss

## Recommitting to Each Other and the Land

TOMMY AKULUKJUK,
NIGORA ERKAEVA, DEREK RASMUSSEN,
& REBECCA A. MARTUSEWICZ

## Introduction

THIS CHAPTER IS A COLLABORATION AMONG 1) AN INUK hunter and artist from Pangnirtung, Nunavut, Canada; 2) a Buddhist activist who has worked in Nunavut for twelve years; 3) a graduate student from Tajikistan; and 4) a Euro-American eco-justice teacher educator, and scholar, granddaughter of a dairy farmer from northern New York State. Here, we each tell a story of love and loss set in very different cultural contexts, where we experienced very contradictory "educations"—formal education, based on job or university preparation, and informal, within our families but based on important ecological values. Our narratives are woven together within an analysis of the historical context of expansionist capitalism and its perpetuations through institutionalized education. Formal state-sponsored education results when we apply Karl Polanyi's *Great Transformation*[1] to teaching and learning. A portion of the vital role of teaching and learning gets extracted from community and delivered over to a new institutionalized process we call "school." The community is de-skilled, Indigenous language is removed, and the plant-animal-spirit-human interrelationship is mocked and discarded as unimportant.

As the stories we tell below demonstrate, responsibility, leadership, community membership, forbearance, humility, and fidelity are the intangible values offered in our families and needed for just and sustainable communities. When these intangible virtues are undermined, the outcomes are quite tangible for all members, human and more-than-human. Our narratives expose the practice of identifying local land-based cultures as "undeveloped" or backward and thus needing improvement via modernization. Schools have been used to accomplish these powerful shifts in cultural priorities and values, moving whole populations off the land to be replaced by extractive machines and systems. And yet, the centuries-old ecological virtues, attenuated as they are by these processes, remain in the cracks and crevices of our day-to-day relationships, practices, and memories. We tell stories to reactivate those virtues here.

TOMMY'S STORY

## *Silatujuq*: Conscious of the Whole Environment and Responding to It Calmly

Modern education has always deemed Indigenous cultures and their forms of knowledge as primitive compared to the regimented and compartmentalized Western system of evaluating student performance. I used to naively think that Inuit, like my father, who can't speak English are not educated, that they had less knowledge of the world than most people. As a child, full of imagination and seeking answers to anything, I thought that my father, too, knew nothing—nothing about the meagre homework that I brought home, nothing about math or social studies. His lifetime of wisdom was deemed not fit for modern education; or, if included, it was labelled "Traditional Inuit knowledge" and relegated to an extracurricular activity, or into a thirty-minute period in a day.

The accountability I have toward my parents gets broken by Western education. The more I go away from my community to get educated, the less I spend time with my family, hunting and learning from Elders. Schools never acknowledge that they are breaking a family continuity that has existed over thousands of years; government can't admit that they are disintegrating Inuit culture. So, I want to try to explain what I think education is, and how it has made me realize the immense knowledge and spirituality that is not taught in schools. How has my father taught me so much about life and taking responsibility for other beings, without ever having to formally teach me? How does love come into the equation of being taught? When does teaching happen?

My father is a very self-composed man. My siblings and I can hardly recollect him being angry at us or at my mother. He is very patient and will sacrifice himself for his family. I can only say this of him now that I am an adult: I never had much view of him while I was growing up, but I knew that he was different from my friends' fathers. He hunted most of his life, and whenever he did work, it was to raise funds for him to go hunting. At the time, I didn't realize how balanced our life was as a family, how much time we spent together camping and hunting. My parents were one of the few parents that insisted that we go camping; they would take us away from school early in the spring, and our summers consisted mostly of camping, which we did until fall. School and Western learning were not practised during these periods, and it was one of the most carefree times of my life. It seems idyllic now.

I point this out because it was the start of my realization of how life might be balanced and how love flowed freely from siblings to parents, and vice versa. It was then that I learned the peculiarities of coming from an Inuit family; after all, it was almost the only thing I knew of life, other than what school taught me. It was a time when perseverance and confidence were built into us, when having compassion for other beings was emphasized, and when we learned that you just had to be nice to everyone. It was a life free of prejudices and discrimination. We all helped in one way or another, every one of us given a chance to complete a task, a chance to catch our first animal. It was a very family-oriented moment, and up to this day, I like to think our family can stick together in the toughest of times.

The traditional Western method of distributing knowledge never acknowledges this love or sense of family. This is a strange thing that only Western cultures can really endure. Why do our most beloved children spend seven hours a day away from their family unit? The only time that children are really inter-acting with parents concerning education is when homework is involved. This can be a very good thing for many families, but when most Inuit come from families where one or both parents don't speak English, it is a rather boring chore. My parents cared about my education; they woke me up and encour-aged me in every way, but school was foreign to them. I can only try to imagine the shame my parents have had to endure not being able to help their own children with something they wanted their children to be good at. (Their per-sistence paid off: out of eight children, six of my siblings, including myself, have gained either a high school diploma or a college certificate.) This was the first time, pretty much in the entirety of Inuit history, that the family unit has been broken by an institution. The sharing of knowledge was broken—a serious consequence, no matter what culture you grew up in.

As Hugh Brody has said, "A hunter gatherer family shares what it has, whether that is information or food. To give to others is to be able to receive from others. Knowledge and food are stored, as it were, by being shared."[2] The sharing that my parents and I had was broken by the English language, which was forced upon us by the institutionalized education system.

This new idea of education to Inuit came with a different sense of being smart, too. Inuit don't really have a word for smart. The closest thing I can think of is *silatujuq*, which can be roughly translated as someone who is conscious about his whole environment and responds to it calmly. He understands his world. I would only reserve this label for Elders and really good hunters, basi-cally someone who has his matters in order. Someone who is smart is someone with the most common sense, and is the most helpful. In some way an enlight-ened person. This word *silatujuq* comes with a serious sense of duty for the community, family, and friends—one will do anything to help, even risk his life. This concept of being *silatujuq* is rarely taught in schools, where smartness is demonstrated by being able to perform tasks well on paper. If you ask me, the concept of being smart in the Inuit way should be taught, or at least tried. Knowledge is only knowledge if you use it for the betterment of your family and community.

How come my father and many Inuit men and women like him, never hav-ing gained a formal education—something that many people pay thousands of dollars for—can be infinitely more knowledgeable about life? To be able to take a situation and assess it properly for everyone's benefit? To still have heavy compassion for people that have hurt you, or to endure situations that command strict self-discipline—I wonder if that is even teachable in our modern school system?

I can tell you from experience that this culture of learning how to take care of oneself is very subtle and you are never told that you are being taught. It is not just your parents who are teaching you, either: you have a whole family of uncles, aunts, and cousins, and they are of all ages. The quality of your learning is only realized when you have mastered or understood what you are supposed to do, when you have a skill developed.

Traditional knowledge, though, is not easily compartmentalized and institutionalized. Inuit traditional knowledge has required its learners to be immersed in the Arctic environment, tied to its happenings and goings. Much of the knowledge is migratory and seasonal: best taught during certain parts of the year. Children were taught to observe very closely the state of furs in the animals, as it indicated when to hunt these animals for their fur warmth.

Inuit children were constantly told to share their catch, which in a way is one of the first steps of serving. Imagine a five-year-old handing out seal meat and being thanked by his or her Elders. From such a young age, we are taught this compassion for our community, and it continues for life. Acquiring skills and knowledge, in the form of hunting and serving your community, is tied to loving your neighbour and your camp (traditionally). But love is a skill that is always neglected in teaching materials nowadays. Modern schools unfortunately are devoid of family love, and the acquiring of skills of different forms of loving is not included in the curriculum.

Among this serving the common good of your camp or community, as rooted in the Inuit culture, lies a series of etiquettes governing relations with people. Inuit children at the earliest age are taught not to talk back to adults, and to only talk when asked questions. (I'm not very old, so I am part of the generation that has been pretentious enough to talk back and not listen at moments.) As Inuit children had to cultivate respect for others, they in turn got the respect that they deserved; individual attention was given them, because it was included with love. This is in contrast to the modern school system, where each social and individual respect is sacrificed in favour of attaining a lot of abstract ideas about social studies, mathematics, and sciences. The modern school system might very well be beneficial, but not when it creates a lot of confusion regarding what our culture is and what we are learning.

Will we survive the onslaught of colonization through the curriculum-soaked Western education system? Will we as members of the earth realize the effects we have on other species? I think this is one area that Western education is really lacking: the attention to our connection to our environments. Western education has taught us very little about taking care of other people and the environment. It teaches singularity and individualism and is rooted in competition. The modern school system believes that there is a top and bottom student, which is very different from the Inuit concept of learning, where every individual is only good when formed into a group or a family. Individuals are to form teams to help members of the community.

Some of what I say might sound grim, or maybe give a sense of being defeated by a system that is foreign, but we have experienced some success at bridging the two cultures. We have Inuit who can converse in both languages and perform well in both types of thought. I don't think this is so much a result of the school system, but a result of Inuit having the support of a family to help us achieve in both systems.

## Learning Respect, Care, and Love from Family and Animals

This story is about ecological and family virtues taught in a relationship with other family members and the environment, and it is told through my personal experiences in my home country, Tajikistan. Situated in Central Asia, Tajikistan became part of the Soviet Union, along with four other Central Asian countries, in the first half part of the twentieth century. When Tsarist Russia and then the Soviet government colonized these areas, they both viewed it as "backward" and in need of industrial modernization because people lived either as nomadic cattle breeders or as sedentary farmers, craftsman, or merchants. Though this lifestyle was more in tune with the needs of the environment, respecting and caring for it, it was not seen as compatible with the requirements of industrial development the Soviet government had planned on. To bring this idea into reality, the Soviet regime saw formal schooling as instrumental to raising a future generation to support industrial modernization. Thus, schooling promoted industrial modernization while emphasizing traditional ways of life as undeveloped and backward. Though this undermined the local relational and ecological virtues that were so integral to the culture in Tajikistan, that culture was not completely lost. Today, regardless of the Soviet Union's urbanization agenda in Central Asia, close to 70 percent of Tajikistan remains rural, and 93 percent of Tajikistan's land is mountainous. Most people still practise agriculture and husbandry as a main source of income. As an agrarian and land-based culture, many of the everyday practices emphasize and pass on these virtues from one generation to the next. These practices, and the virtues they held within my family and community, shaped my understanding and appreciation of the relations we built together. One of my early memories of learning ecological and relational virtues was experienced when I visited my grandma.

I grew up in a city called Khujand, but during my summer break I used to visit my grandmother in her village, where I spent three months a year helping her with house chores. The village was located where water came from the mountain. Crops like wheat, tobacco, potato, corn, grapevines, and many other types were grown on the hills. And people built a pond to store drinking water. Because no one had a tap to drink from, we carried water in buckets for drinking, cooking, and general household needs. People shared that pond, which meant every family took care of it with the strict agreement to not pollute it. The pond where we took water to drink would be replaced two to three times per month by a stream coming from a nearby mountain. One day my grandma took me to the pond and showed me how to fill the bucket with water. But before showing me how to do it she said that I should be very careful not to drop any dirt into it, as the village—including us, of course—used it for drinking and other household activities. She told me about a little boy from another house who peed into it. The whole village asked his father to stream a new source of water, which he did. That was one of my lessons about the importance of water in our life. I learned that water was the source of life; if we pollute it or don't take care of it, we harm ourselves.

My grandma had a busy life; she had land, farm animals, and a big household to take care of. I was given different chores, from taking care of the house to feeding domestic animals. Through these responsibilities I remember learning about the ecological and relational virtues that were part of our village life. We grew crops like wheat, corn, and vegetables, and over the table I would hear my grandma talk about her concerns when the water stream was low and how it was going to impact crops in the village. She talked about sharing the stream with neighbours, making sure that every family gets their share of water. That was a big part of village life, where people came together to share its bounties for the common good. One day she asked me to fetch water for the cow and calf. I said it was too hot outside for me to catch water for them. She explained to me that these animals were an important part of our life in the village as they give us produce, and in return we must take care of them. They needed to drink water to make more milk and gain strength. By feeding and watering them three times a day, I grew to love them as friends—indeed, as part of my family. We practised love and care for each other and it made a whole difference.

I also learned to clean the barn and take out and dry its manure. When I felt disgusted by its smell, my cousin noticed this and said that though it smelled bad, once it was dry it would be burned to heat the house in the winter and to cook food in the summer. And the manure was used to fertilize the land that fed the cows and us as well. I learned to appreciate its importance in our life and how everything was connected. Nothing went to waste. By living these direct relationships with the land, water, and animals, I learned about the concept of interdependence. These relationships taught me ecological and relational virtues that would have escaped me if we had relied on industrialized agriculture as the only way of life. They were place-based and experiential: I knew we needed each other and it could not be otherwise. This intangible knowledge and wisdom was not taught at school but experienced in the family. I learned that whatever wrong I did had an immediate impact on me, and I developed a profound knowledge about the world around me.

Being a part of the family, everyone had a responsibility to contribute to our well-being. These responsibilities were carried collectively with family members and neighbours. Through these responsibilities I learned to respect and care for family members, neighbours, elders, and younger community members. I learned the importance of sharing. We shared land, water, and labour with neighbours and other villagers. One of the vibrant memories I have involves the harvest season. There was corn, wheat, and tobacco; and neighbours would help each other as everyone would take turns doing the work. Harvest was done manually with farm tools. Once neighbour A got done with harvesting, everyone moved to help neighbour B. During the tobacco harvest, we used to go to the field in the morning and harvest ten to twenty sacks of tobacco to bring home. Once we were home those sacks of tobacco would be divided between each family group, who would put it through the needle to hang and dry it. We would do this work sitting in a big circle, sharing jokes, stories, and getting comparative and playful with each other. I remember working on it a whole day and feeling such a sense of accomplishment afterword. We also came together to help each other at big gatherings like weddings, celebrations, or funerals. A close net of relationships and membership was developed in these ceremonies, which strengthened communal life by teaching us interdependence and appreciation of each other. Once we knew each other through these intimate moments, it was harder for us to hurt, undermine, and disrespect each other.

These land-based cultures were big on communal life; it was never about "me," it was about "us." I remember having a conversation about not putting myself above others. These conversations came as we learned to do work in a group and finish tasks collaboratively. Every morning after breakfast my cousins of similar age and I would carry a bucket of water and fill all the big containers to be used by the whole household. We would spend at least forty-five minutes filling all big containers with multiple visits to the stream-fed pond. Though I felt really tired each time carrying water, I knew I was doing my part of the responsibilities in that household. This communal life also taught me to recognize that whatever food was put on the table was possible thanks to the hard work of community members coming together. That is why it was a vice to waste food or disrespect something that nourished our body. We were taught to appreciate our meal every day and the bounty of the land that was so generous to us. At the end of a big meal, we recognized through prayer the hard work of the people who took care of the crops. We knew it was hard work because we were part of harvesting. Through these kinds of conversations and prayers we were taught to respect the land and care for it. These moments experienced and lived in relationships, care, and sharing were not always smooth. But it was the core of my upbringing. I loved my life in the village because I could run, play, and be outside the whole day. Watching TV or being on any electronic devices was unheard of during my life in the village. Instead, I was in direct contact with a close-knit community and learned crucial skills to sustain myself. School taught me many things, but it was never a replacement for the wisdom and humbleness that teaches us we are part of a bigger ecological system.

Another crucial life lesson was taught to me by my mother. She would always spend time to explain things and emphasize the importance of humility and sharing. To keep me focused on my chores, she would remind me to keep helping my grandmother and that that was one form of respect and care we could show her. She taught me to show respect to grandparents and elders and be courteous and kind to neighbours. She worked full-time as a teacher, but she always made sure to make time for long conversations about humility, kindness, care, and respect. From her I learned to remain open to things beyond my understanding and to value the complexity of life. She taught me what it is to be just, to question things when they are not right, and to always seek answers. This helped me to question the industrial and technological forces that were portrayed as better replacements for agrarian life. With her wisdom, kindness, sharing, and care, she taught me by example how to live life. Thanks to her I learned to understand community needs and build relationships with others as a crucial means of well-being and happiness. Everything she and my grandmother spoke of, their teachings, helped me find my place as a member of both the human and more-than-human world.

Growing up, I do not think I understood any of these teachings in depth, but now I realize that they are the main pillars of my understanding of the world. I understand now that we need an education that teaches us the humility to recognize ourselves as a part of a bigger ecological system upon which we depend emotionally, physically, economically, and socially; education through which we learn that we are one member of a complex web of life, and that our responsibility is to take care of this complex web for the common good.

REBECCA'S STORY

## "There's Nothing Left Here Anymore"

My story echoes many of the themes raised in my Inuit and Tajik colleagues' narratives, but from a very different cultural context, that of a small rural town in northern New York State. While I recognize the critical historical and cultural differences in our life stories—I am a direct descendant of the devastation perpetrated by white European settlers[3]—there are connections both in the land-based value systems learned as children and the systematic undoing by expansionist capitalism of those patterns of relationship that protected social and ecological well-being in our communities.

My first and arguably deepest education was informal, made of lessons learned primarily from my mother; the second came from school. My childhood at home was filled with lessons of kindness and caring for other creatures (horses, dogs, cats, rabbits, trees, plants, fields), respect for elders, knowledge of local woodlands, fields, and streams, and responsibility to my family and community through orderliness and hard work: cultivating and preparing food, tending animals, caring for children, keeping house. From the back of a horse in my backyard, I was taught patience, self-discipline, compassion, embodied inter-species communication, and collaboration. While she refused to see them as generating important knowledge, these values and skills were brought to my siblings and me from my mothers' experiences growing up on a small dairy farm just a mile from our house.

My grandparents established that farm in the early 1900s, when about 40 percent of the US population worked on small farms. Grandpa milked around sixty Holstein cows, and when my mother was growing up also raised chickens, hogs, and champion Percheron horses. The horses were used to plow the fields and do other tasks around the farm that needed such strength as they could provide. My mother claims that she was raised by those gentle giants, having lost her mother to breast cancer when just a toddler: "The housekeeper used to get me out of her hair by putting me up on the back of one of Dad's big horses and slapping it on the rump to send us out into one of the fields. More than once I found myself stuck in the grain bin, unable to get off! I spent a lot of hours like that!"

Besides being funny to imagine, those hours atop the back of a draft horse taught my mother, even as a very young child, patience, humility, and gratitude. She grew to love horses, dogs, and all sorts of other creatures like they were best friends. And she learned the land for miles and miles around the farm and the village from the back of her pony. Forget a bicycle; she had a horse! She could tell us who lived in what farm, what condition the land was in, who worked for the farmer, and what kind of crops were grown there. Of my grandfather's farm, she understood traditional crop rotation and the use of cow and green manure to fertilize fields. We used to drive by, smelling the fresh manure on the breeze, and she'd comment to us about which crops—corn, alfalfa, winter wheat—were in which fields and why. Grandpa used those fields to graze his cows or to grow the crops needed to feed them, selling hay to neighbours or folks in the community who had a horse or two if there was any to spare. Our milk came from the farm in a small milk can that Grandpa would carry in the back door along with a creel full of rainbow trout he'd caught that afternoon in the stream across from his house.

My mother would cook the lean little trout, heads and all, served with vegetables from her own garden or one up the road belonging to a neighbour. She knew how to cultivate, harvest, cook, and can from a huge garden on the side lot next to our house. She taught me how to weed, find the ripest fruit, and shuck peas. And she knew the best places to find blackberries, from which she made jams and pies—my father's favourite. She was thrifty, practical, and brilliant when it came to pie pastry, horse sense, wildflowers, and a whole host of other day-to-day skills that made our lives together healthy. And it was all based on a particular form of love, for us and for the land and other creatures.

Eventually, all this practical knowledge came to me, though I didn't know it as knowledge (or even love) at all, and neither did she. She did not claim those skills as knowledge; in fact, she did her best to deny their value, ironically, even as she insisted we learn. Farmers' kids (or grandkids) were "hicks." My mother felt that especially sharply. By the time I was growing up, in the 1970s, the number of people on farms in the United States had shrunk to just 4 percent. This decline was accomplished via four systematic methods,[4] the effects of which I witnessed and experienced in my own extended family: 1) USDA agricultural policy began to push mechanization and the use of credit to encourage farmers to buy bigger and bigger tractors and other equipment, as well as chemically based fertilizers and pesticides that created financial and nutritional dependencies and ultimately wore out the soil; 2) the use of government subsidies to encourage a "get big or get out" approach to farming, which also preached big technology and export crops over diversified and self-sufficient systems that could support families and the community; 3) the rise of a degrading cultural discourse and class stratification system defining farmers and their families as backward, uneducated "hicks"; 4) patterns within formal schooling that defined success in terms of monetary earnings gained by "experts" in professions that could be found anywhere *but* those small rural towns like the one I grew up in.[5]

My mother once told me how happy she was to be "rescued" from the farm by marrying my father. She only moved a mile away, but she "got out." It broke my heart when she told me this. This exclamation, and my own story one generation later of being schooled (and encouraged by my parents) to find happiness by leaving my town to "get out," "become educated," and "make something of myself," reflect these broader historical and cultural patterns. I could never imagine myself staying in that little town, though now I deeply regret that response and even envy my classmates who stayed to work and raise families there.

From its beginning in the establishment of that farm, my family's story is part of the development of what Wendell Berry calls the "unsettling of America"[6] and the dislocation of land-based cultures across the world by expansionist capitalism. We are learning from critical scholars the effects of settler colonialism on Indigenous Peoples, especially as these are reproduced in schools.[7] Extending this work, we should also study the ways the violent extractive processes of late capitalism have unsettled the settlers, impoverishing the people and the land they once tended.[8] The social and ecological effects are profound, as my once thriving rural town (with others across the United States and the world) is now dying: rivers polluted by CAFO run-off, topsoil depleted, local grocers, clothing stores, hardware stores, diners, and cafes replaced by Walmart, Lowes, and McDonalds. Prisons are now the number one employer in St. Lawrence County. As my Aunt Mary, the last remaining family elder, said to me recently, "Of course the kids are leaving, there's nothing left here anymore."

But if it's not too much to say, can we discern a common thread among the activities and efforts that these small towns turn to, to rally, resist, and resuscitate? Might that thread be love? Can't we see in our own lives many instances where the appeal of a flash fire of money can't compare to the warmth of friendliness, helpfulness, co-operation, and care? How do we move beyond the profit-driven gluttony that pushes people off the land and impoverishes all, if not by turning toward the bonds of love?

## Cease to Do Evil (Then Learn to Do Good)

There is a pithy saying by the Buddha that goes, "Cease to do evil, learn to do good, purify the mind. This is the way of the awakened ones." It's the order that impresses me. A lot of first-world activists think our job is to "rescue" those suffering from theft and bullying by our corporations and military. And a lot of first-world meditators want to rush to the last order of business: getting our busy minds purified. But isn't the first job to cease to do evil? Especially when most of the evil on this planet is being done in service of our land-thieving, fossil-fuel-burning way of life? Living in the belly of the beast, we are likely the best positioned to stop evil from being done; Claire Culhane convinced me of this.

In the 1960s, Claire had been a nurse in Vietnam, where she watched her hospital get (illegally) used as a military base by US-supported South Vietnamese troops. When she came home, she discovered Canada was the single biggest outside supplier of weapons used by the United States in Vietnam. So Claire began a protest movement and wrote several influential books before being profiled in *One Woman Army: The Life of Claire Culhane.*[9] Claire frequently quoted Che Guevera's observation: "I envy you North Americans, you are very lucky. You are fighting the most important fight of all. You live in the belly of the beast."

My turning point with Claire was a conversation at a Burger King in Ottawa. She told me about finishing her term as a nurse in South Vietnam and going to Paris to meet the ambassador from North Vietnam. She asked him if she could go volunteer as a nurse to help the North Vietnamese. He politely refused her: "Miss Culhane, let the bombs fall on our heads. Why don't you go home to Canada and stop the bombs from being built in the first place?" Her advice echoed the original teaching of the Buddha, and it motivated me to do much of what I've done since. Why was I so receptive to her advice? I think the answer lies in the ethical foundation of intercultural respect, valour, and trust conveyed by my brother, my parents, grandparents, and mentors.

Courage. The only person who has ever physically intervened to defend me in a fist fight is my younger brother. Once, when I got cornered and beaten up by the school bully, Grant, who was very small in stature and only eight years old, jumped in and ferociously attacked the much older and bigger guy. That still stands as one of the bravest things I've seen, and lives on in my brother's attitude of commitment, self-sacrifice, and incandescent love.

Respect. Our parents encouraged us to study and appreciate different cultures. They pushed me to study world religions in high school, giving me my first exposure to Buddhism, which now plays such a big role in my life.

Justice. My father volunteered to fight in the Second World War, as did my grandfather and various uncles, and my mother went through the London Blitz as a child. With this background, I was deeply affected by reading *The Crime and Punishment of IG Farben* (1978) in my final year of high school. Nuremberg war crimes prosecutor Joseph Borkin documented how major corporations like AGFA, BASF, Bayer, and Standard Oil (now Exxon) profited from running factories attached to concentration camps. That book lit a fuse in me—one that burned faster when set to a punk music soundtrack of DEVO, Elvis Costello, and the Clash.

Activism. When I finally met Professor Bill Phelan in my first week at university, I was primed for a shift. Bill became my principal mentor. I was enrolled in business administration, but as soon as I heard Bill describe his Sociology 110 course as an investigation into the "world emergency" posed by US militarism and the threat of first-strike nuclear war, I promptly signed up. To this day, I still pass out Bill's reading list: Chomsky, Wallerstein, Seymour Hersh, Gabriel Kolko, Dorothy Dinnerstein, Shulamith Firestone, Ivan Illich, etc. Those readings rocked my world.

Within a year I would be engaging in civil disobedience to block a cruise missile factory outside Toronto, and helping to found the Alliance for Non-Violent Action. Also in that period, Julia McCoy and I founded the first East Timor group in Canada, through which we met Noam Chomsky when he agreed to come up and do a speaking tour about Timor at Canadian universities. After four or five civil disobedience arrests, we had our first trial, where the inspirational activist Father Philip Berrigan came to testify for our defence. But all these arrests started to bring up a fundamental fear: I was terrified of being locked up for a long stretch with just my own mind for company.

So I began to look for a meditation teacher, and by default leaned toward the Buddhists I'd read about in high school. My activist heart was with the Catholics and Quakers, but they didn't seem to have any methods to help you mentally survive incarceration. Buddhists seemed to, but on social justice matters they were—for the most part—annoyingly silent. My mentor Bill suggested I go to meet Namgyal Rinpoche, a Canadian who had been a Communist Youth member and a monk in Burma before being recognized by the 16th Karmapa and the 14th Dalai Lama as a reincarnated teacher in the Tibetan tradition. When I told Rinpoche I was a peace activist, he said: "Excellent! Keep it up." A year later I signed up for a three-year training program for ministers that Rinpoche had founded.

A year after graduating from that, I went to Nunavut for what I thought would be three months; I stayed for twelve years. Much of what I write about today comes from what I heard or read over those years from Elisapi Ootoova, Mariano Apilardjuk, Annie Quirke, Zebedee Nungak, Kenojoak Ashevak, John Amagoalik, Monica Ittusardjuat, Malaya Nakasuk, Jerry Ell, Joe Kunuk, Joanasie Akumalik, Tommy Akulukjuk, David Joanasie, Aluki Kotierk, and many others.

Watching the dominant civilization—my people, the *Qallunaat* (Euro-Americans)—bully, hound, and harm Inuit and other civilizations is painful. But to watch our civilization run around the world trying to "rescue" the people that we pushed overboard to start with is outrageous. As I understand the virtues espoused by my family, Claire, Bill, and others, the work placed in front of us first worlders is to stop our civilization from brutalizing and stealing from others. We should cease to do evil before we presume to teach anyone how to do good.

## Extractivism and Western Education

Colonialism and capitalism are based on extracting and assimilating. My land is seen as a resource. My relatives in the plant and animal worlds are seen as resources. My culture and knowledge is a resource. My body is a resource and my children are a resource because they have the potential to grow, maintain, and uphold the extraction-assimilation system. The act of extraction removes all of the relationships that give whatever is being extracted meaning. Extracting is taking, stealing, taking without consent, without thought, care, or even knowledge of the impacts that extraction has on the other living things in that environment. Colonialism has always extracted the Indigenous—extraction of Indigenous knowledge, Indigenous women, Indigenous Peoples. Children from parents. Children from families. Children from the land. Every part of our culture that is seemingly useful to the extractivist mindset gets extracted. There's an intellectual extraction, a cognitive extraction, as well as a physical one.[10]

Leanne Betasamosake Simpson notes the parallels between the extraction of resources and the extraction of knowledge. Her insights may help explain why the single biggest new spending item in Canada's 2014 right-wing federal budget was $1.9 billion for First Nations education.[11] To some, this might have looked like a contradiction: Why would a Harper government keen on the tar sands pour money into schools on reserves? In fact, the two go hand in hand. People are fond of the forests and rocks that the economy feeds on. If humans still draw their primary sense of identity from a place, then it's difficult to move and allocate them to where the economy needs them. Roots—and all those virtues that keep us rooted—get in the way. Hence the economic utility of changing from a rooted form of socialization among a "tribe" in "place" to a system with less solidity and less solidarity. Institutionalized Western education dissolves land loyalty. You have to extract the people from the land if you want to extract the stuff under their feet. This is the same basic story that Rebecca tells about her experiences in northern New York. Small farmers tend to be very connected to the land and the communities that they serve. To make dairy farming a business, it was necessary to convince the next generation that it was dirty, ugly work, and to school them (us) away from the community. Agricultural extraction, as Berry tells us, meant creating the means for mono-crop agricultural processes in order to shift from land-based relationships that fed communities to industrial processes that pulled crops, animal bodies, and profit out of communities and into large corporations as a matter of "efficiency." To do so, farmers, convinced of the need to "modernize" (often via land grant universities' agricultural programs) are forced to go into massive debt to banks, as well as seed and machinery companies. They are told that they are "feeding the world" and that this is a Green Revolution, but it's anything but "green" and their families find themselves stretched so thin that one or both spouses often has to go off-farm to work.

Inuit call people who behave like this *Qallunaat* due to "their materialistic nature," says Mini Aodla Freeman: "The word implies humans who pamper or fuss with nature. Of materialistic habit. Avaricious people."[12]

When *Qallunaat* cannot persuade Indigenous Peoples to enclose their places and permit extractivism, then we unleash institutional Western education on them. This mode of institutionalized education extracts and encloses teaching and learning. Enclosure of the land is a mission of capitalism, enclosure of learning is the mission of Western education.

Excavating rooted Indigenous learning and persuading communities to replace it with a placeless education service is the key to undermining their sense of belonging and concern for the land. Before companies and states can take the stuff under a people's feet, they have to convince them to embrace placeless learning. In the United States and Canada, this has been going on since at least the early nineteenth century, but has since been exported across the world as an essential tool of "development." When people "learn from the land," then they aren't too happy to see it trashed. But if they learn in a school—what do they need the land for? What is developed, then, is a powerful psychological alienation from, and material enclosure of, the interdependencies needed for life.

# Antidotes: Local Virtues

The world's tyrants... (have a) weakness.... They have no knowledge of the surrounding earth. Furthermore, they dismiss such knowledge as superficial, not profound. Only extracted resources count. They cannot listen to the earth. On the ground they are blind. In the local they are lost.... Effective acts of sustained resistance will be embedded in the local, near and far.[13]

One antidote to these trends noted by John Berger is to identify and critique them as we have done here. To point out their non-universality, and the taken-for-granted assumptions behind them. Colonialism and its later incarnations (capitalism and neo-liberalism) have destroyed the core of the cultural matrix that generates the kind of virtues that support earth communities.

Against the fragmentation caused by mechanistic ways of knowing and the reductionist imaginaries and policies they incur, Wendell Berry insists on a recognition of the diverse and unifying connections that tie us to each other and with the living world—bodies, minds, and spirits—as the mysterious source of all possible ways of knowing or being. "These things that appear to be distinct are nevertheless caught in a network of mutual dependence and influence that is the substantiation of their unity. Body, soul (or mind or spirit), community and world are all susceptible to each other's influence... each part is connected to every part."[14] Such, we argue, is love the necessary basis for life. We recognize the bonds of love in our own family histories, in the relationships and practices that we experienced as children and by which we learned to be responsible for one another. We now work to revitalize this responsibility in our own teaching and learning.

Another antidote is to slow down and build locally. We are not as far gone as capitalist elites want us to believe. We can reweave the fabric of community. Paul Goodman said, "Suppose you had the revolution you are talking and dreaming about. Suppose your side had won, and you had the kind of society that you wanted. How would you live, you personally, in that society? Start living that way now!"[15] Goodman is asking us: What are the virtues we want to embody and bring to life? Now go do that. When we wake up tomorrow morning, how will we act in accordance with this? The community garden that we cultivate, the choir we sing with, the house we've helped our friends build, the animals and forests we have learned to care for, the time we spend caring for a sick relative—all these activities pass Goodman's test. Think of the ways we cultivate rootedness and responsibility to landscapes and to each other and foster joy and creativity. Now go do them!

We believe that virtues are soaked up through the lived experiences of people (families and communities) living together in place, working and playing together, and struggling together, and that it is through hearing about people's narratives of such lived experience that we learn best about the virtues.

We must begin to listen to Indigenous wisdom holders who have been pointing out the harmfulness of *Qallunaat* ways for centuries. And we must tell our own stories, where the needed virtues still pop up through the cracks in the system.

Our stories are of what we have learned through relationships: with animals, as Tommy and Rebecca and Nigora describe; with our elders, with our families, and with our mentors and teachers. We have developed strong, principled love for these people and relationships, principles that guide our willingness to protect what we love in these places and beyond. For us, it's the best way to fight back against the violence engulfing the world. Generosity, humility, fortitude, kindness, determination, and forgiveness: we bring these virtues to life by becoming the teachers and mentors, spouses and parents, friends and activists that we wish to see in the world. Together, we defend these virtues, resist wrongdoing, and build the good.

Importantly, we do not act alone. Che Guevara said: "The desire to sacrifice an entire lifetime to the noblest of ideals serves no purpose if one works alone." Friendships and mutual admiration spawned the collaboration in this chapter and indeed in this book. It would take hours to list the ways in which each of us has been encouraged and inspired and tangibly supported by the others here to persevere in our activism, scholarship, mentoring, and community-building. These bonds of respect and nurturance are the lifeblood of our defence of the places, people, land, and relationships that we love and that have loved us into being. As Joanasie Akumalik says: "Qungapassi" (I smile at you all).

## Notes

1 K. Polanyi, *The Great Transformation: The Political and Economic Origins of Our Time* (Boston: Beacon Press, 1957).

2   H. Brody, *The Other Side of Eden: Hunters, Farmers and the Shaping of the World* (Vancouver: Douglas & McIntyre, 2001), 193.

3   E. Tuck and W. Yang, "Decolonization Is Not a Metaphor," *Decolonization: Indigeneity, Education and Society* 1, no. 1 (2012): 1–40; L. Veracini, *Settler Colonialism: An Overview* (New York: Palgrave MacMillan, 2010).

4   W. Jackson, *Consulting the Genius of the Place: An Ecological Approach to a New Agriculture* (Berkeley, CA: Counterpoint, 2011).

5   W. Berry, *The Unsettling of America: Culture and Agriculture* (San Francisco: Sierra Club, 1996).

6   Berry, *The Unsettling of America.*

7   Tuck and Yang, "Decolonization Is Not a Metaphor"; E. Tuck, M. McKenzie, and K. McCoy, "Land Education: Indigenous, Post-Colonial, and Decolonizing Perspectives on Place and Environmental Education Research," *Environmental Education Research,* 20, no. 1 (2014): 1–23; D. Calderon, "Uncovering Settler Grammars in Curriculum," *Educational Studies* 50, no. 4 (2014): 313–38.

8   R.A. Martusewicz, *A Pedagogy of Responsibility: Wendell Berry for Ecojustice Education* (New York: Routledge, 2018).

9   C. Culhane, *Barred from Prison* (Vancouver: Arsenal Pulp Press, 1981); Culhane, *Still Barred from Prison* (Montreal: Black Rose Books, 1985); Culhane, *Why Is Canada in Vietnam? The Truth about Our Foreign Aid* (Toronto: NC Press, 1972); M. Lowe, *One Woman Army: The Life of Claire Culhane* (Toronto: Macmillan Canada, 1992).

10  L. Betasamosake Simpson, in N. Klein, "Dancing the World into Being: A Conversation with Idle No More's Leanne Simpson," *Yes!*, March 6, 2013, http://www.yesmagazine.org/peace-justice/dancing-the-world-into-being-a-conversation-with-idle-no-more-leanne-simpson.

11  J. Rae, "Behind the Numbers: Harper's New Funding of the First Nations Education Act," Olthuis, Kleer, Townshend LLP, February 14, 2014, http://www.oktlaw.com/behind-numbers-harpers-new-funding-first-nations-education-act/.

12  M.A. Freeman, *Life among the Qallunaat* (1978; reissued, Winnipeg: University of Manitoba Press, 2015), 86.

13  J. Berger, *Landscapes: John Berger on Art* (London: Verso, 2016), 251.

14  Berry, *The Unsettling of America,* 110.

15  Goodman cited in R. Solnit, *Hope in the Dark* (Chicago: Haymarket Books, 2016), xxiv.

# THIRTEEN

# Evoking Ethos

## A Poetic Love Note to Place

CARL LEGGO & MARGARET MCKEON

## Margaret

CAREFUL AND ARTFUL ATTENDING TO OUR PARTICULAR places holds strong medicine for our times. As humans, we are both created by, and co-creators of, the world. We create the world through our being and doing and in our words, our languages, our stories. In this time of weighty ecological messaging and trauma, environmental discourse is often a very heavy thing—sometimes it feels too heavy to consider lifting, carrying, or shifting. Elin Kelsey and Rick Kool show how environmental education, with its "focus on loss and threats,"[1] does not offer supports with which to bear the attendant weight of despair and hopelessness. We put forward that to imagine a changed reality of loving, responsible being, we must first speak our world, in all its magical and terrifying complexity, as if it is a place of love. We are made of the places where we live and those we come from. We are made of their histories. Where we have been is the first foundation of any imaginal place of more ethical or virtuous living we might seek. As Keith Basso tells us, "we *are*, in sense, the place-worlds we imagine."[2] Just as small streams gather into rivers that then enter the ocean, we write poetically to weave fragmentations of time and place into imperfect wholeness, to write ourselves into ethical being. Poetry is a virtue that lingers in intimacy to know a broader world complexly. To know dark winter and spring, death and beginnings as balanced rhythms— in lives and environmental movements both. Through sometimes painful, sometimes joyful inquiry, in telling our personal and communal stories, we affirm relationships of past, present, and future. As poets, we undertake this task of world-creation, of this love note to place, as a sacred task. We invite you to join us in the tumbles and turns of our entwined stories as we ourselves entered them, perhaps as a floating leaf allows for the mysterious unfolding of flowing water.

# Carl

In the process of writing poetry we slow down and linger with memories, experiences, reflections, and emotions. We need poetry because poets engage with what Ted T. Aoki calls "playful singing in the midst of life."[3] Poets are always attending to language—including the alphabet, grammar, diction, syntax, spelling, connotation, music, and imagery—as well as the keen intersections of the mind, heart, body, spirit, imagination, and memory. As Jane Hirshfield claims, poetry brings "new spiritual and emotional and ethical understandings, new ways of seeing, new tools of knowledge."[4] In poetry we seek new ways of knowing, being, becoming, doing, and hoping. Parini wisely claims that "poetry offers an antidote to the bludgeoning loud voices of mass culture, insisting on the still, small voice, the personal voice, thus staking a claim for what used to be called the individual soul."[5] We need to hear our own voices and the voices of others, singing out with hopeful courage and conscientious commitment. As poets, we seek to live in the places of language, emotion, and imagination, and wonder, always convinced that poetry calls out in vibrant voices—activist, creative, transformative, and passionate. We compose stories that nurture composure for living well, conscious of the connections that sustain us creatively in the creation.

## Margaret and Carl

We have both lived for many years in the same neighbourhood in Corner Brook, a city of twenty thousand people, located on the west coast of Newfoundland in the Bay of Islands, a sub-basin of the Gulf of St. Lawrence. Corner Brook is nestled in the Humber Arm, one of many inlets in the Bay of Islands, surrounded by the Long Range Mountains. While Corner Brook is isolated—a day's drive to St. John's, and an even longer journey to what Newfoundlanders still call the Mainland of Canada—it is home to the Corner Brook Pulp and Paper Mill, once the world's biggest paper mill, and the Grenfell Campus of Memorial University, with its thriving arts community. It is also the administrative headquarters of the Qalipu First Nation band government, a recently established Mi'kmaw Nation. Corner Brook is a place that is always changing and evolving. Though we have since moved away, we are steeped in this particular place. We are variously rooted through social and natural relationalities. While the stories we tell about ourselves are always unique and coloured by the keenly experienced sense of individual selfhood and subjectivity, our seemingly unique stories are inextricably connected to many other people and the communities that help inform and shape our sense of identity and purpose. Susan Griffin understands that "the self does not exist in isolation" because "to know the self is to enter a social process."[6] We agree with Griffin that identity is "less an assertion of independence than an experience of interdependence,"[7] and, therefore, "for each of us, as for every community, village, tribe, nation, the story we tell ourselves is crucial to who we are, who we are becoming."[8] In this chapter we evoke ethos and complex histories and explore how to live well by engaging together in a braiding of poetry, stories, and ruminations that attend to our nurturing of loving relationships in our personal, pedagogical, prophetic, and political commitments. We are committed to the ecological virtue of love, especially as expressed in hopes for sustaining well-being through consciousness, composure, and connection.

## Carl

March 2017. As I do about twice a year, I am visiting Corner Brook. Everywhere I look I see snow. Memories of childhood swirl around me as fast as the falling snow. Adam Gopnik calls winter "the white page on which we write our hearts."[9] I have returned home where poems fall from the sky with the swirling snow. David A. Greenwood notes that "poetry, like music, leads me back home, not merely to ideas about life, but to the embodied experience of living."[10] I have returned home in order to walk and snowshoe in winter, to know winter again intimately with/in the body and imagination. I write poetry as a way to connect with place, people, experiences, and emotions, a way to attend consciously to the swirl of each day's complex busyness. I write out of love. I am eager to remember, and even more eager to attend to the momentous mystery of the present, full of wonder.

### Snow

mesmerizes hypnotizes dazzles
enthralls enchants entrances

a swirling vortex maelstrom whirlpool
pandemonium turning in turbulent turmoil

I want to surrender to snow disorder
disarray like the unseen face of God.

snow shimmers subdued harmonic
a waltz in a grand Venetian castle hall

luxuriant carnivalesque exuberant
wild erotic dangerous exorbitant

my heart cannot hold winter
will likely burst with an aortic aneurysm

snow burns with fierce sensuality
snow rewrites the story constantly

the sky might be falling
into a creation conjured in wildness

the more I look the more I see
the more I see the more I look

# Margaret

I write from Vancouver, where daffodils and magnolias are each day smiling more broadly, trees' leaves are beginning their slow stretch of renewal toward the sun, and snow is a dream held by the surrounding mountains. Having moved here over a year ago to study, I am beginning to get my feet under me. Still, a larger part of my heart remains as knit round with Newfoundland wool as the snowbound parking metres on Broadway in Corner Brook. In late March in Corner Brook, snow-winter still holds strong and the living earth rests under banks, mounds, piles, and drifts of snow.

In Coast Salish Vancouver, on the Pacific Coast, I must adjust my rhythm to be in a new place. I remember that every place is its own wholeness and one not a lesser or greater version of another. I am able to become present as I stop troubling comparisons and let my heartbeat adjust to the tides of a new ocean. To let myself celebrate that moment when leaves on trees say *now*, now is the moment to push through this tight sheath of winter pregnancy.

My body has travelled away from my heart and identity. I am stitching myself slowly back together through surrender to the rhythms of beach stones, people, and birds renewing nests. I remember this lesson as I attend to discourses of environmental loss and threat. In all the environmental change that knits through my head and heart, crows are collecting sticks, anticipating lengthening light and new camouflaging leaves. They are anticipating the tender raising of new young. I remember to lie still against tree trunks and watch closely the birds at work; to be wise in a place is always to start all over because this is wisdom that isn't mine.

*Here, Whatever the Weather*
*after Al Pittman's* Lupins

Future tellers no longer predict
a world changing so fast now.

Their grace swallowed
in earth, her new cracks and tremors.

Still snowflakes follow one-after
unhesitant of streets' lit threshold.
The tide sureness of them.

Still tracks trail boots, trace fingerprints
of black on icing sugar streets.
The solidness of them.

Tonight too unrelenting shadows
crispest that whipcrack moment each flake
marries the ground, distinguished.

We scattered lupin seeds for you
in October, not the August of package directions.
What certainty has there ever been
of the miracles of soil under snow?

# Carl

In "The Road Taken," Dwight Garner asks: "How do we create a new home for ourselves without forgetting the place that will always be home?"[11] At sixty-three years old, I have now lived one-half of my life in Newfoundland, and one-half more in four other provinces of Canada, especially in British Columbia, where I have lived since 1990. I have written several books of poetry and a novel about growing up in Newfoundland and about leaving and returning home. In Come-By-Chance,[12] I wrote particularly about the experience of diaspora, about leaving the East Coast of Canada for the West Coast of Canada. My poetry could be interpreted as an expression of nostalgia or homesickness, but that is not accurate. I write about home, not for nostalgic sentimentality, but for gnostic sensitivity. In other words, I am not so much homesick as I am eager to know the experience of home. I do not want to forget where I came from. If we are going to know who we are, we must know where we came from. I seek ways to nurture connection, to invite and sustain the kind of dialogue that shares intimate stories, raises tough questions, and evokes new possibilities for living with love. In all my writing I seek to remember the past so as to live well in the present. Like Barbara Kingsolver, "my way of finding a place in this world is to write one. This work is less about making a living, really, than about finding a way to be alive."[13] And as Méira Cook understands, "there are places that cannot be paraphrased."[14] Perhaps those places are especially the places we call home. Christina Baldwin claims that "story is how we come home,"[15] and certainly I have written many poems and stories in my efforts to go home and know home.

### Rooted

I grew up in a small house
nailed to the side of a hill
in a milltown on an island
anchored in the North Atlantic

I have lived in many places
since, but when my friend met
me at the airport in Whitehorse,
I said, Bob, wherever I go

I am reminded of Newfoundland,
and always the diligent psychologist,
Bob said, That's rather egocentric,
and I said, No, I think it's geocentric

I am always rooted in a place.
Wherever I go I always remember
where I grew up like a love affair
lingers in the heart's remembering

## Margaret

I was twenty-one when I first arrived in Corner Brook. It was a sunny and cloudy day in September 1999. Laden with my belongings for the school year, my long walk from the bus to the boarding house was interrupted by an insistent offer of a ride. "I saw you walking there a half-hour back," she said. "You don't want to climb Elizabeth Street." And she sure was right; the city map had nothing to say about steep streets. Even in those early years, I don't remember feeling homesick for Edmonton, where I grew up. And yet leaving Newfoundland for PhD studies fifteen years later wrung my heart out like a sponge. I wasn't sure if I knew who I was without the place. The intensity of my place-longing was woven through with uncertainty of my ever having belonged—an illegitimacy of my rootedness. In selling the furniture to rent my home, one man paused with a filing cabinet we were carrying down the stairs to say, "You're not from here, are you," evidently pleased he had puzzled this out of my language, gesture, presence. I laughed. This identity as a *come from away*, a *mainlander*, didn't concern me until I was living elsewhere. At the same time, friends and neighbours who live up and down my bay-looking hill ask me, "When are you coming home? Home to stay." "When will you be back on Fudges Road for good?"

My dad grew up in an Irish Catholic neighbourhood on Staten Island, in New York City. Like his grandmother who lived in their family home, much of the community was comprised of recent immigrants from Ireland, though as he remembers it, no one ever spoke of Ireland. Indeed, in travelling to Ireland myself a few years ago, it felt like a piece of me that had been severed was given back. The land was so *loud* to me. My return to North America was with a new living skin. I wonder about the enduring unnamed longings and imperfect belongings of the diaspora when I hear environmentalists assert that the environment would be better off without "us." That humanity is a blight on the planet. Never, as Potawatomi scholar Robin Wall Kimmerer relates of sharing with incredulous ecology students, that the land might both love and need us. That we might belong on earth.[16]

### Cod Whispers

I do terribly at first try with that knife; fillet of cod
is a thing to buy fresh or frozen, her severed spine
the stiff back of this place I've never touched.

Her big dish eye, the full moon's tugging linger
over water's slow dawn, fingers of craggy cliffs
and the morning-raising calls of birds.

Her tight scales, the shimmering surface
of the bloody flesh-deep and ancient dreams
foreign to a beef bucket in a smeared double kayak.

Her slippery fight, a gut memory in small boats
that a long food fishery this year is political, that gulls
flock over boatloads of austerity-axed libraries.

You sever her spine-length with spiritual precision
but it's me leaving the Island, me with ragged flesh,
fins strung on the wrong side of a bloody knife.

I dream to be more blueberry, her gentle offerings
yield softly to fingertips and jam on toast, reached
surely by a worn footprint in adjacent juniper.

Cupped in my hands, I move a mosquito hawk,
all gangle and fight, from my kitchen to dewed grass,
but it's possible too to be whispered out willingly.

# Carl

In all my writing about growing up in a little house located at 7 Lynch's Lane,
Corner Brook, Newfoundland, I seek to attend to the particular place, to
remember it, to research its impact on my living, to honour the people who
have dwelled and continue to dwell there. Like Greenwood, I ask, "How do
we pay attention and what do we pay attention to?"[17] I am committed to paying attention, especially in poetry, in order to know where I am in relation to
others, and who I am in relation to others, and how I am in relation to others.
Dionne Brand suggests that "places and those who inhabit them are indeed
fictions."[18] She also recommends that "to be a fiction" is to be "in search of its

most resonant metaphor."[19] What is the resonant metaphor I am searching for? In all my writing I am filled with desire for understanding who I am, how I am, where I am, especially in relation to the creation. When I was growing up in Corner Brook, my father often took my brother and me on long hikes to ponds hidden in spruce, fir, and alders where we fished for trout. We always packed a big lunch for the long day. My father taught us from the beginning: "Leave nothing behind. Always take your garbage home." My father taught me as a boy that we had a responsibility to live respectfully with the creation. We enjoyed the wilderness; we enjoyed fishing for trout; we lived in relationship to the land. I have lived my life with a keen sense of gratitude for the places where I have lingered. I do not want to take these places for granted; I do not want to forget them; I want to honour the gifts they have granted me.

Brand claims that "landing is what people in the Diaspora do."[20] My first book of poetry was titled *Growing Up Perpendicular on the Side of a Hill*.[21] Lynch's Lane, a steep, narrow gravel lane that was always a challenge to drive in any season but especially treacherous in the winter and spring. In the winter snowplows and big trucks delivering oil for heating homes were almost the only vehicles that ventured down Lynch's Lane. Only pedestrians climbed the lane in winter. Spring always washed the lane out. Lynch's Lane was so steep that two-story houses were jammed into the side of the hill so the front was two stories and the back was one story. The house I grew up in was a trapezoid. The rail fence ran up the hill like a lopsided ladder.

### Lynch's Lane

like black lines burned in wood
with a glass for focusing
the sun in a point
Lynch's Lane is etched in my body

the first orange popsicle
later lime grape pineapple even
but none ever tasted as good
as the first orange popsicle
of summer with mosquitoes

sweat stinging sunburn
water and tar on the lane
to keep dust down
Skipper mowing the grass
with whistles of the scythe

autumn potatoes no bigger
than jumbo marbles boiled
in the skins sprinkled
with salt the world afire
in squashberry crumbles bakeapple jam
blueberry pies partridgeberry jelly
the wind rustling restlessly
with Cec, Frazer, Macky,
my brother, and me playing war
cricket kick the can at day's end

sucking icicles knocked from the eaves
hot cocoa drunk over the furnace grate
Old Mrs. Eaton climbing Lynch's Lane
grasping the fence like a ladder
picket after picket gasping through wool
noses pressed to frosted windows
homemade bread with Good Luck margarine
howling winter winds the house like an ark
mothers calling, Where are you going?
You can't see your noses

orange sherbet dipped in chocolate
the pink flesh of fried trout
all the neighbours in their yards
shovelling snow searching for crocuses
fallen leaves holding the sun
and cutting shapes in ice
everywhere the air lemon
smell of freshly washed cotton

the world melting splashing washing
away like saints in the River Jordan

like black lines pricked in skin
with a needle for focusing
India ink in a point
Lynch's Lane is tattooed in my body

# Margaret

For a long time, I lived and I taught outdoor and environmental education as if the land did not already have stories—as if the trees, animals, and mountains were *terra nullius*, a blank slate for my and our experiences. For our activism. But as much as it is important for each new generation to experience and imagine its worlds anew, the land and the possibility (and impossibility) of our physical presences in our places are deeply storied. Each stretch of mountaintop, Prairie river valley, and coastline on this continent is part of an Indigenous nation's traditional territory, with languages and cultural traditions "as different from one another as are other nations and cultures in the world"[22]—as varied as the landscape itself. Within the Haudenosaunee context, Joe Sheridan and Mohawk-Haudenosaunee Dan Longboat describe the inseparability of land and its Indigenous Peoples, how this sacred language contains the true names of things themselves, and how through long dialogue between land and culture "old-growth minds and cultures mature, emerge, and encompass the old growth of their traditional territory."[23]

Each house and property, each national park trail, each continuous connecting mile of highway holds its own story of, often violent and almost always deceitful, dispossession. This conquering of land and its First Peoples is not just of the past, but continues—it will not be long now that the illuminating lamps of my house in Corner Brook will be powered by the new Lower Churchill Dam in Labrador. I consider my complicity in the flooded place-stories and gravesites, in this stilled sacred river, in the mercury-poisoned fish that vibrate the wires of my home.

## Teaching with Corner Brook Mi'kmaq
### for Kevin

ERASURE: Mi'kmaq of emerging Corner Brook
barred from the new Mill and its British town
Poverty, rocks and slurs thrown at the young.
Mothers raise their children well with invented
pirate grandparents. Safely outside themselves.

CLOSURE: Beothuk starved, diseased and shot.
Hat on heart, we loudly sing the sorrow, build
on Beothuk Crescent, camp at Beothuk RV Park.
Stonewall DNA testing of Beothuk perseverance.

The lies I taught had been scrubbed clean
long before I arrived to the Island. Witness
the Mi'kmaw coming apart of clenched fists.
Broken river ice jams the shore: RESURGENCE

Ancestors haunt me as a bear-drum warms
the stonework at city hall. They call at dusk
through deep mountain snow, face me
when I face the children. My work is not choice.

You teach our thirsty students, "It doesn't matter
the plastic in their back pocket. It's how they walk
with respect on Mother Earth." And you love me,
even as I bloody my face on glass walls of whiteness.

I am not Mi'kmaw, but while we work
I learn to forgive myself. To find my spirit
in my own life's stories, to bind my feet
to this land's stories. To *Ktaqmkuk*.[24]

We pray, *M'sit nogama*.[25] Might it be, "The land
knows you even when you are lost."[26] Teach me

how shame is a white cane not a leaden pack,
how loving relationship, sight through dense brush.

# Carl

I am diasporic. I left Lynch's Lane for the first time when I was fifteen. I went to Quebec on a student exchange. In lots of ways, I have always been on the move. I am always dispersing. Even on Lynch's Lane in those years of seasonal cycles and few other changes, in the midst of familiarity, I often felt like I did not belong. I was too bookish, too scholastic, too serious! But perhaps we are all diasporic. Perhaps we all feel like we do not belong. Like Linda Hogan, I think that "emptiness and estrangement are deep wounds, strongly felt in the present time."[27] Perhaps we are always writing fictions about places and our relationships to places so we will feel rooted, physically located, psychically connected, imaginatively and emotionally anchored, even if the ground is always a shifting tectonic plate or template. Ernst Bloch asks, "How do we ever know who we are,"[28] and he offers what sounds like a dire warning: "life then and life now have no connection, or merely one in melancholy."[29] I do not know why Bloch thinks the only connection between the past and the present is melancholy. With the etymological link to the Old French *melancolie* (sadness, gloom, anger), melancholy refers to an ancient medical understanding of an excess of black bile in the spleen. Melancholy might certainly be a possible response to the past, but hardly the only one. I seek a more hopeful vision of the past, including even the sad, gloomy, angry stories of the past. I recognize the past as past, but also present in the present. In my return to Corner Brook and Lynch's Lane, I am still learning to land. I like Lorna Crozier's observation: "As I get older I realize more and more that you don't lose anything from the past." And, then, she adds: "We shouldn't forget where we come from."[30] That is the abiding motivation of so much of my writing. Regardless of where I live, I do not forget where I have come from. I do not erase the drafts of my life stories. Like a palimpsest, the drafts and traces are always still present. As are the lessons learned from my father. In many of my poems, I call my father "Skipper," an honorific for older men often used in Newfoundland. From my father I learned many lessons about love, especially about living in connection with people and places. I learned about creation as whole and wonderful, as holy and hopeful. My father taught me to seek composure as a way to sustain well-being.

### Homework

When my son was young,
most nights I helped him
with his homework
and remembered how
Skipper sat close beside me
on the edge of his big bed
while I memorized textbooks
and answered questions.
In spite of long days
in the mill and frequent calls
from neighbours to fix their
ovens, toasters, electric kettles,
Skipper always quizzed me
for tests, sometimes for hours,
and never complained.
When we studied
geography, Skipper said,
*Wherever you go, know*
*where you come from*
*so you can find your way back.*

## Margaret

I miss winter. There is no winter in Vancouver—not a proper winter and certainly not a Western Newfoundland winter. In Corner Brook, by late March, a walking trail surface might be two feet higher than an adjacent plowed lot and one must be careful to not trip on the park bench tops. I love how, come spring, the long, hidden earth is like a forgotten friend returned from travel. Its slow revealing is familiar yet profoundly new. There is a return of the unfinished stories of fall: a lost hat or garden tool or a garbage can lid flung by a fall wind that is unveiled in a yard up the hill.

Winter, especially in big snowstorms, insists to us humans that we are not in charge. The world spins on other axes than our clocks and is written with other rules than our property lines and roads. In his work on eco-psychology, Andy Fisher describes, particularly as seen in the lives of trauma survivors, interpersonal reactions "which often swing dramatically between states of intense attachment and terrified withdrawal."[31] I wonder how this kind of trauma lens might lend understanding to Western societies' simultaneous intense attachment to land control and ownership and their disavowal of land relationship. What are the complex entwinements of relationship with land and the enduring, unresolved traumas of North America's diasporic peoples? In my homesickness for Newfoundland, I am endeavouring to learn how this pain can help teach me to hold land and place more lightly.

### Another March Storm

I'm tired. My cat in the window follows my moving,
removing, from one claimed place to another
that which keeps falling. Blows back where it was.
I'm bailing a holey hull but the ship never sinks.

Each day is a new day with leaning towers
to fashion tulips if I wanted or a beach.

+++

My co-worker walked off his roof last night.
Just like that. Walked off and on into the house.

+++

Weather is a masked robber, makes off
with Routine. Angry! Violated!
Time was robbed as if never ours.
We're squirrels defending nest eggs.

But work must begin at 8:06 each day.
Control is water, oxygen, a food group.

+++

The white unhappiness is not outside.
It swirls around our bodies, gathers
in all the public places. And my office.
Nowhere is safe.

+++

Snowshoe in woods, double bent
and hollow with quiet. Wallow
in hip deep power. Rest. Laugh.
Remember animals with better paws.

# Carl

In winter in Newfoundland snow falls and falls. It renders the world like a sheet
of white paper, an invitation to write more stories, but those stories will not
be the final stories. The snow will melt in the spring, and the tracks and traces
in the snow will disappear, but the stories will have been composed and they
will linger. In her tender and thoughtful memoir *Belonging: Home Away from
Home*, Isabel Huggan writes about the "carnal knowledge of landscape."[32] This
is always my experience when I return to Corner Brook. I know the Western
Newfoundland geography and landscape in my body with sensual awareness
and indelible memory. In *View from My Mother's House*,[33] I write affectionately
about the world I saw from the vantage of the house on Lynch's Lane where I
grew up. I especially liked to watch the Humber Arm, a long inlet in the Bay
of Islands that begins in the Gulf of St. Lawrence. Cargo ships sailed into the
Humber Arm frequently. In the winter an icebreaker often carved a path for the
ships. When I was a boy, neighbours sometimes skated across the frozen bay.

The house on Lynch's Lane is long gone. Decades ago, most of the houses
on Lynch's Lane were torn down so a road could be constructed through the

neighbourhood. My parents moved a few kilometres away to a lovely, flat cul-de-sac on the Humber Heights. I have no regrets about the destruction and loss of Lynch's Lane. It was a challenging place to live, and my parents' stories in another house have been full of joy. It is ironic that in recent years a developer has constructed a housing complex in the area. With an ambitious imagination, the developer dumped tons of rock and gravel in order to create a flat foundation for houses. Much of the rock and gravel was dumped in what was the backyard of the old house at 7 Lynch's Lane. People now pay a great deal of money to own one of the new duplexes with a view of the remnants of Lynch's Lane and the Humber Arm, which never changes and, yet, is never the same on any two days. While growing up in Corner Brook, I did not fear change. Every year, season, month, and day brought changes, but I always understood how everything and everybody was connected. Only after leaving Corner Brook did I begin to grow fearful. Linda Hogan claims that "we have been wounded by a dominating culture that has feared and hated the natural world, has not listened to the voice of the land, has not believed in the inner worlds of human dreaming and intuition."[34] In my poetry I seek to reclaim voices and languages and wisdom I once knew in Corner Brook, so I can live with more love and wellness in other places.

## Winter Alphabet

returning in March after years
of Vancouver winter rain

I know only I have forgotten
the winters I grew up with

for a few days I walk in Corner Brook
as if I am fighting winter

head down, going somewhere fast
except I move slowly

almost pantomime, pushing myself
through winter like walking under water

I must learn to lean with winter
seek its erratic rhythms

like a dory sliding up and down
the smooth sides of a rough sea

I taste winter, winter savours
my body with a lustful lover's appetite

snow bites pinches pokes stabs
slices like a set of sharp knives

in a TV infomercial
neatly skinning a tomato

snow acts with verb exuberance,
a veritable thesaurus of action words

winter reduces the world
people stay home more

huddle in their cars more
hide in shopping malls more

deep snow, hard-packed snow,
plowed snow, powder snow

no hint of spring anywhere
except spring always comes

sunglasses essential, blind colour,
light and shadow tear the retina

snow in mountain creases
and cracks, a monochrome world

like the alphabet on paper,
a text I am learning to read again

reminded how quickly I grew
illiterate, lost my language

# Margaret

There are moments in my co-teaching of sustainability projects that stand out as important stories to hold and carry forward. There was the day in a grade 5 class that we were making a web on the board of all the economic, cultural, and non-human connections to the Newfoundland fishery that we could think of. Near the end of our process, one student, after adding "pollution," wisely insisted that "pollution has to be connected to everything else." And, gesturing with her arms, "If there is too much pollution, there is nothing left, nothing!" No songs or stories, jobs, transportation networks, fish quotas, or seafood grocery counters. No seaweed gestation for the young cod or capelin to fertilize roadside gardens. Her gesture didn't hold just the web at the front of the room, it swept across the classroom window, through which the mountain-brimmed Bay of Islands stretched out in winter light.

Another day we were learning about the 1992 cod moratorium, which closed the five-hundred-year-old fishery that was the economic, cultural, and social mainstay of most all Newfoundland rural communities. Teachers began to share their memories of starting university alongside men undergoing "retraining" and of the large-scale physical movement of families. Though their world is crafted by it, many students hadn't heard of the moratorium. Students went home and asked for memories from their families and social communities and later created art pieces to share back with the community.

David Smith helps me understand the dynamics of continuity and rupture in community change. Through his hermeneutical lens, in which "the question of continuity has assumed a status of vital importance," he groups together a number of change paradigms that, though they seem diverse, all "work from a vision of radical rupture and separation rather than conversation."[35] In another class on forestry, I was conscious to not dismiss the origins of students' families who, for generations, had been employed in extractive forest industries. I believe that the seeds of most change are to be nurtured from within us and our histories. These seeds were within the students' own experiences of sadness and loss on the land. Story, in its various forms, weaves continuity through and into change.

### Walk with the Ghosts that Travel Old Humber Road
### for Carl Leggo

They've sprouted condos among the ghosts
that walk Old Humber Road. From grass-wild slopes,
old godbless-scrawled foundations, rogue lilacs.

Ghosts' footsteps flow like water, undeterred on a trail
ever tracking where road should be. Through
an October clay-slick pit of sewerpipe replacement.
Past the unheeded sign *hard-hat-only, construction.*
Around a king-of-the-castle snowplow pile.

Of your storied lane, I know only that as February
turns March, Orion past dusk perches
over the Humber Heights' backdrop ridge.
That an eagle hovers in huntsman's place by day.

That the florid antihero hill your boy-feet careened
this Sunday morning is quiet. Quiet but for the sole-crunch
of gritty road ice. Gulls that slice frozen steam snorting
from the mill. Snow curtaining in over the Humber Arm's
mouth-mountains, trucks on sea ice again
and the flat taste of unsalted breeze.

From your poems, past stories settle a fresh snow
on my walk and walk with me. Whip about in wind.

In the last long-grass yard of the last standing
squat woodbox house, skipper breeds cars' parts

and piled pallet scraps. He carves a just-so trail
that contours the snow moraine blocking the road,
services sand. Tomorrow, shovels it again, gently.
carving into the ever-backhoed craggy rise.

Condo paint dry, the well dressed tether good cars
on asphalt driveways and the blue casts of big screens.

They and we-the-footsteps all: interlopers of home.

# Carl

When my father was dying, in 2008, I often walked the hallway of the palliative care wing while he held my arm. On one wall was a black and white photograph of a winter night on West Street taken in the era when the shopping districts of Corner Brook were West Street and Broadway. Since the cars parked on the side of one-way West Street were about the same age as my father's old 1963 Bel Air, I assume I was about ten years old when the photo was snapped. Snow is falling under the streetlights. The photo is peaceful, still, silent—appropriately sacred for a palliative care wing. Perhaps I recall that photo because I saw it in the heightened emotional space of waiting for my father to die, or perhaps I saw it as a lasting memory of childhood, long past but still indelible.

Dionne Brand notes that "snow is quiet. It is not like rain. It has the sound of nothing happening. It is like a deep breath held and held."[36] As I waited for my father to die, I held my breath, and my father's breath held me. It was autumn, and the sun shone with a frigid finger like an angry teacher just before lunch. The photo on the wall held the past—the memory of snow light was frozen, a trace of light from a distant dead star. Now, almost a decade after my father's death, I still remember the photo on the wall of the palliative care unit. It is steeped with memories and emotions. As Greenwood observes, "everything is related to everything else."[37] I know almost nothing about the photo. I do not know the name of the photographer, or who framed the photo and hung it on the wall of the palliative care unit, or when it was placed there. All I know is that in the heightened emotional space of walking with my father as he faded away with a brain tumour, my memories were sharpened and intensified. Above all, my memories were infused with love. And that is how I remember Lynch's Lane, where I grew up with my father, mother, brother, sister, grandmother, and a network of extended family and neighbours. As Patrick Lane knows, "Much can be learned from a patch of forest or an open meadow. Much can be learned from a backyard."[38]

### Trembling Aspens

the forest presses heavy
on the low light, full
of hope for names

shadows everywhere
a dark counterpoint
in the light spaces
of air fire earth water

drawing silence
like the sun calls the sea

light and shadow are
the letters of the alphabet
rendering spaces visible
so we can see

even as language dissolves
revolves solves involves
resolves meaning deferred
deferentially in difference

still clutching the wild
chaotic world in my words

trembling aspens whistle
the wind with winter's end

# Margaret

We should not underestimate the risks of opening ourselves to feel the more-than-human world as if it were an extension of our own skin. To love a tree is to mourn when it is cut down—for paper or new tennis courts—or when it falls in the wind of an autumn storm. To appreciate the routines of birds is to feel the ripples in the interconnections that wind around that single tree.

For a long time environmental education offered issues-based approaches that assumed that people's actions would change if only they knew the particulars of environmental distress. As a young nature guide at a zoo, I was expected to ask three-year-olds who were enamoured with the tactile softness of a wolf pelt to take up the plight of wild wolves. David Sobel describes a numbing condition that such an approach risks, where the "overwhelmingness of environmental problems can breed a sense of ennui and helplessness."[39] Andy Fisher writes of how we each already bear pain from the distress of natural worlds, even if we are not awake to this underlying source that stretches across the landscapes that surround us. He criticizes approaches that use and rely on us to be beings burdened with anger, fear, and shame to provoke change.[40] As with other traumas that we are asked to face in our lives, I believe it is vital to first weave safe and loving containers for the work. These containers help us be strong and resilient, and to be clear in who we are so we can face the struggles of the world with our best grace. These are containers woven of the love found in both human and more-than-human relationship. They are built from being embedded in and knowing place. From working with Mi'kmaw cultural teachers, I began to understand change as a call to responsibility through and within loving relationship. Tewa scholar Gregory Cajete writes about community as "the medium and the message"[41] of Indigenous education and of being human. Community, encompassing the human and more-than-human, comes from long and continually renewing commitment to relationships—that is, to love.

## Be We to the Good

Be we to the good, like a Newfoundland man
(if he's salty enough) to a car off the road.
Like urbanites to their workaday press
or some places' strangers to fear.

*I wouldn't want to live* THERE, you say,
(by some freeway in the States) where
no one would stop or check a car
wheels-up in the ditch. *Right*, I nod.

Down the bay-gaping pitch of my hill a pickup
is side-perched up a high March snowbank.
His driver door pinned, corn-snow asphalt
glares back. Then a swarm of trucks, ten
Newfoundland men touting shovel-rope-winch.
Clear. Before my feet find boots.

Snow a thick screen on a dark-tunneled highway.
I'm the lone car through a hollow hour
of untracked-nothing till two trucks cloud past.
I'll meet them. Two bends past a rolled car,
I ease to turn back. But there's trucks flashing and men
windbent running, flashlights knifing snow
from the dark. I drive them to the car,
its tag and telling footprints, graciously empty.

On alone through drifts, salt sifts through my mind:
salt for roads but in waters, wind. Inhaled breath.
What we would be, if, without attachment or fear,
we could—be for the good. Not just for roads
but caring, through all colours of loved lives.

## Carl

In a biography of Harlan Hubbard, the Kentucky painter and writer Wendell Berry explains that Hubbard "wanted to know the earth in its particulars, for that is the only way that it can be known with love."[42] He adds that "Harlan never forgot that people's relation to the earth was inescapably economic as well as aesthetic and spiritual."[43] This is my understanding of my many connections to Lynch's Lane. Corner Brook was built around a pulp and paper mill that was for many years the biggest paper-making mill in the world. The paper from the Corner Brook mill was shipped to the big daily newspapers in Boston and New York. When I was growing up, you could often smell the sulphur in the air. An almost constant stream of smoke, steam, and fumes poured out of the tall stacks like long, lean fingers holding a cigarillo high in the air. Nobody knows what kind of effluent was poured into the Humber Arm. I only know our parents warned us not to go swimming anywhere near the mill (and, uncharacteristically, we obeyed our parents). In the midst of corporate and industrial imperative and pollution, I learned as a boy that I was related to the earth, connected to the earth, responsible to the earth as the earth was responsible to me. On our Saturday trips to hidden ponds for trout fishing, my father taught me to leave no garbage. My father's advice was a metaphor for living with reverence wherever we lingered. My father was a good man, a part of a generation of working-class men who were glad they had enough money to pay the bills and enough energy left on days off to help one another. My father loved Corner Brook and he loved Lynch's Lane as he loved Curling, where he grew up, a few kilometres away. He travelled very little in his life. Like Harlan Hubbard, he "wanted to know the earth in its particulars, for that is the only way that it can be known with love." I have travelled far more than my father, and I will almost certainly continue to live far away from Lynch's Lane, Corner Brook. I now know Steveston, British Columbia, intimately and gratefully as home, but I learned to live with a poet's heart in Newfoundland, and I will always remember my father's wisdom:

> *Wherever you go, know*
> *where you come from*
> *so you can find your way back.*

## Surprise

in Corner Brook,
20 degrees below zero,
the harbour is frozen,
snow is piled high
as the STOP sign, still
falling, cold and windy

all week, the sky, overcast
forever, except today
a ray of sunshine hints
at spring's promise,
no fire, a candle flame,
hungry and hopeful

# Margaret

I wanted to mark the passing of the spring equinox. It had been a long week. This morning I took my prayers down to the ocean, for wider and nearer worlds and for myself. I offered my skin to be touched by the icy water, by ancient memories and songs of rebirths.

I sat for a long time, the waves in and out, until I felt stilled. I was still so long a young gull began to browse the receding tide near my feet.

*Arise my love, arise. Arise into your work. The day is still strong. Follow your hands and your heart into whatever warrior work the universe has waiting for you. It's for the poetry of your being alone.*

### Make Me a Flower

Through road sand come crocuses. I too
after winter underground

uncostume a flower. Let me be moved
by smallest wind, water,
sunrise; keep your angry stories.

Give me phosphorescence of coral
and thrum of spring birdsong. Like a solar still,
collector of joy in the desert.

Run brooks through me,
heaviness of oil decompose at my feet.

## Notes

1   Elin Kelsey and Rick Kool, "Dealing with Despair: The Psychological Implications of Environmental Issues," paper presented at the Third World Environmental Education Congress, Turin, Italy, October 2005, 2, https://pdfs.semanticscholar.org/860d/85e6f28fc14e91a38308b67fa763cb15a4f3.pdf.

2   Keith H. Basso, *Wisdom Sits in Places: Landscape and Language among the Western Apache* (Albuquerque: University of New Mexico Press, 1996), 7, emphasis in original.

3   Ted T. Aoki, *Curriculum in a New Key: The Collected Works of Ted T. Aoki*, ed. W.F. Pinar and R.L. Irwin (Mahwah, NJ: Lawrence Erlbaum, 2005), 282.

4   Jane Hirshfield, *Nine Gates: Entering the Mind of Poetry* (New York: Harper Perennial, 1997), 79.

5   Jay Parini, *Why Poetry Matters* (New Haven, CT: Yale University Press, 2008), 63.

6   Susan Griffin, *The Eros of Everyday Life: Essays on Ecology, Gender and Society* (New York: Doubleday, 1995), 50, 51.

7   Griffin, *The Eros of Everyday Life*, 91.

8   Griffin, *The Eros of Everyday Life*, 152.

9   Adam Gopnik, *Winter: Five Windows on the Season* (Toronto: House of Anansi, 2011), 215–16.

10  David A. Greenwood, "Creative Tensions in Place-Conscious Learning: A Triptych," *Journal of the Canadian Association for Curriculum Studies* 13, no. 2 (2016): 12.

11  Dwight Garner, "The Road Taken," *Esquire*, February 2017, 52.

12  Carl Leggo, *Come-By-Chance* (St. John's, NL: Breakwater Books, 2006).

13  Barbara Kingsolver, *Small Wonder: Essays* (New York: HarperCollins, 2002), 233.

14  Méira Cook, *Slovenly Love* (London, ON: Brick Books, 2003), 34.

15  Christina Baldwin, *Storycatcher: Making Sense of Our Lives through the Power and Practice of Story* (Novato, CA: New World Library, 2005), 224.

16  Robin Wall Kimmerer, *Braiding Sweetgrass: Indigenous Wisdom, Scientific Knowledge, and the Teachings of Plants* (Minneapolis: Milkweed Editions, 2014).

17  Greenwood, "Creative Tensions," 15.

18  Dionne Brand, *A Map to the Door of No Return: Notes to Belonging* (Toronto: Doubleday Canada, 2001), 18.

19  Brand, *A Map to the Door of No Return*, 19.

20  Brand, *A Map to the Door of No Return*, 150.

21  Carl Leggo, *Growing Up Perpendicular on the Side of a Hill* (St. John's, NL: Killick Press, 1994).

22  John Borrows, *Recovering Canada: The Resurgence of Indigenous Law* (Toronto: University of Toronto Press, 2002), 3.

23 Joe Sheridan and Dan Longboat, "The Haudenosaunee Imagination and the Ecology of the Sacred," *Space and Culture* 9, no. 4 (2006): 366.

24 Ktaqmkuk is "Newfoundland" in Mi'kmaw.

25 M'sit nogama is "all my relations" in Mi'kmaw.

26 Kimmerer, *Braiding Sweetgrass*, 36

27 Linda Hogan, *Dwellings: A Spiritual History of the Living World* (New York: W.W. Norton, 1995), 84.

28 Ernst Bloch, *Traces*, trans. A.A. Nassar (Stanford, CA: Stanford University Press, 2006), 27.

29 Bloch, *Traces*, 62.

30 Lorna Crozier, "Seeing Distance: Lorna Crozier's Art of Paradox," in *Where the Words Come From: Canadian Poets in Conversation*, ed. Tim Bowling, 139–58 (Roberts Creek, BC: Nightwood Editions, 2002), 142.

31 Andy Fisher, *Radical Ecopsychology: Psychology in the Service of Life*, 2nd ed. (Albany: SUNY Press, 2013), 98.

32 Isabel Huggan, *Belonging: Home Away from Home* (Toronto: Vintage Canada, 2004), 4.

33 Carl Leggo, *View from My Mother's House* (St. John's, NL: Killick Press, 1999).

34 Hogan, *Dwellings*, 84.

35 David G. Smith, "Hermeneutic Inquiry: Imagination and the Pedagogic Text," in *Forms of Curriculum Inquiry*, ed. Edmund Short, 187–209 (Albany: SUNY Press, 1991), 194.

36 Brand, *A Map to the Door of No Return*, 145.

37 Greenwood, "Creative Tensions," 11.

38 Patrick Lane, *There Is a Season: A Memoir* (Toronto: McClelland and Stewart, 2004), 41.

39 David Sobel, *Childhood and Nature: Design Principles for Educators* (Portland, ME: Stenhouse, 2008), 146.

40 Fisher, *Radical Ecopsychology*.

41 Gregory Cajete, *Indigenous Community: Rekindling the Teachings of the Seventh Fire* (St. Paul, MN: Living Justice Press, 2015), xiii.

42 Wendell Berry, *Harlan Hubbard: Life and Work* (New York: Pantheon, 1990), 32.

43 Berry, *Harlan Hubbard*, 52.

# Author Biographies

**Tommy Akulukjuk** is a hunter, artist, and writer from Pangnirtung, Nunavut. A graduate of the Nunavut Sivuniksavut Training Program in Ottawa, he has worked for the Lands division of the Qikiqtani Inuit Association and as an environmental researcher for the national Inuit group Inuit Tapiriit Kanatami. He has also advised the territorial government on establishing a new Inuit Cultural School for Nunavut.

**Heesoon Bai**, PhD, a recipient of the Excellence in Teaching Award and the Dean of Graduate Studies Award for Excellence in Supervision at Simon Fraser University in Canada, researches and writes about the intersections of ethics, ecological world views, contemplative ways, Asian philosophies, and psychotherapy. Her SFU profile is available at http://www.sfu.ca/education/faculty-profiles/hbai.html.

**David Chang**, PhD candidate, is a teacher and teacher educator in the Faculty of Education at Simon Fraser University in Canada. David taught secondary English for a decade before working as a faculty associate with Professional Programs at SFU. He studies ecological ethics, sustainable communities, contemplative practices, and ecological ways of life.

**Douglas E. Christie** is professor of theological studies at Loyola Marymount University, where he teaches in the area of Christian spirituality. His primary research interests focus on contemplative traditions of thought and practice in ancient Christian monasticism, on spirituality and ecology, and most recently on traditions of spiritual darkness and unknowing and their capacity to help us respond meaningfully to the contemporary sense of exile, loss, and emptiness.

**Paul Crowe** is an associate professor and chair of the Department of Humanities at Simon Fraser University, where he also teaches for the Asia-Canada Program and directed the David See Chai Lam Centre for International Communication from 2008 to 2015. His research and publications have addressed classical and contemporary fields of inquiry. His SFU profile is available at https://www.sfu.ca/humanities/people/faculty.html.

**Nigora Erkaeva** is a PhD candidate at Eastern Michigan University. Her dissertation employs a critical discourse analysis of the introduction of compulsory education and historical change in Central Asia in the early twentieth century.

**Thomas Falkenberg** is professor in the Faculty of Education at the University of Manitoba, Canada. He is the editor and co-editor of five books, including *Sustainable Well-Being: Concepts, Issues, Perspectives, and Educational Practices* and the recently published *Handbook of Canadian Research in Initial Teacher Education.* More details about his research and academic background are available at http://thomasfalkenberg.ca.e.

**David Greenwood**, professor in Graduate and Undergraduate Studies and Research in Education at Lakehead University, is Canada Research Chair in Environmental Education. His publication list is available at https://www.lakeheadu.ca/users/G/dgreenwo/node/17468.

**Mike Hannis** is a senior lecturer in ethics, politics, and environment in the School of Humanities at Bath Spa University, UK. He is also a member of the Research Centre for Environmental Humanities at the university. From 2014 to 2019, his research in environmental ethics formed part of the interdisciplinary *Future Pasts* project (see www.futurepasts.net) funded by the UK Arts and Humanities Research Council.

**David W. Jardine** has recently retired from a full professorship of education. He now works with groups of teachers in the Calgary, Alberta, area in attempting to decode the strange circumstances of schooling and to cultivate more sane ways of taking up the task entrusted to teachers and students in schools.

**Peter H. Kahn, Jr.**, professor at the University of Washington with joint appointments in the Department of Psychology and the School of Environmental and Forest Sciences, is director of the Human Interaction with Nature and Technological Systems Laboratory. His faculty profile is available at https://faculty.washington.edu/pkahn/.

**Dr. Carl Leggo** was a poet and professor in the Department of Language and Literacy Education at the University of British Columbia. As an arts-based education researcher, he was part of the a/r/tography movement at UBC. His UBC profile is available at http://educ.ubc.ca/professor-carl-leggo/. For some of his voluminous publications, check https://www.researchgate.net/profile/carl_leggo.

**David R. Loy** is a professor of Buddhist and comparative philosophy and a teacher in the Sanbo Zen tradition of Japanese Zen Buddhism. He lectures nationally and internationally on various topics, focusing primarily on the encounter between Buddhism and modernity and what each can learn from the other. He is especially concerned about social and ecological issues. For more, visit www.davidloy.org.

**Rebecca A. Martusewicz** is a professor in the Department of Teacher Education at Eastern Michigan University. She recently published *Pedagogies of Responsibility: Wendell Berry for EcoJustice Education* (2019). Her faculty profile is available at https://www.emich.edu/coe/departments/teacher-education/faculty/martusewicz-rebecca.php.

**Darcy Mathews** is an assistant professor in environmental studies at University of Victoria. He is an ethnoecologist and archaeologist and works in collaboration with First Nations communities to understand the deep history of social and ecological relationships between past peoples and their environments.

**Margaret McKee** is currently an associate professor of social work at Lakehead University in northwestern Ontario, Canada. Marg's PhD is in counselling psychology, but she is also a former obstetrical nurse and professional musician.

**Margaret McKeon** is an outdoor educator, poet, and doctoral candidate in language and literacy education at the University of British Columbia. In her research, she, a person of Euro-settler ancestry, considers land relationship, ancestral knowledges, and colonialism through life writing and poetry.

**Derek Rasmussen** is an activist and meditation teacher trained in the Burmese and Tibetan Buddhist traditions. He also works as an advisor to Inuit organizations on education and social issues. Currently, he is a PhD candidate in the Faculty of Education at Simon Fraser University, writing his dissertation on the teaching of Buddhist meditations on unbounded love.

**Charles Scott** teaches at Simon Fraser University in the Faculty of Education as an adjunct professor and at City University in Canada (Vancouver) in the School of Education as an associate professor.

**Nancy J. Turner**, CM, OBC, PhD, FRSC, FLS, an ethnobotanist and ethnoecologist whose research focuses on traditional knowledge systems and traditional land and resource management systems of Indigenous Peoples of western Canada, is distinguished professor emeritus in the School of Environmental Studies, University of Victoria, Canada.

**Jan Zwicky** is the author of nearly twenty books of poetry and prose, including *Songs for Relinquishing the Earth*, *The Long Walk*, and, most recently, *The Experience of Meaning*. Zwicky grew up in the northwest corner of the Great Central Plain on Treaty 6 territory, was educated at the Universities of Calgary and Toronto, and currently lives in a coastal rainforest succession on Canada's West Coast, unceded territory with a complex history, including Coast Salish and Kwakwaka'wakw influences.

# Index

conservation, 8, 22, 215, 229n6
consumerism
  and advertising, 192–93
  Buddhist perspectives on, 124–25
  and the contemplative gaze, 149
  and the ecological crisis, 125
  and economic participation, 217
  and happiness, 94
the contemplative gaze
  overview of, 137–38
  in Christianity, 145
  and consumerism, 149
  Czeslaw Milosz and, 143
  John Muir and, 148–49
  and moral failures, 138
  in papal encyclicals, 139–40
  poetic examples of, 143–44
  power of, 140
  rarity of, 149
  as spiritual-ethical practice, 147
  and strange affinities, 147–48
  Thoreau and, 147–8
contemplative practice. *See also* meditation
  and attention, 63–64
  benefits of, 64
  in Buddhism, 127
  in Christianity, 145
  confronting climate change, 63–64
  and cosmic consciousness, 145–46
  and creativity, 151
  cultivating, 147–48
  and ecosattva, 117
  and ethical commitments, 145, 150
  the present moment, 146–47, 151
  significance of, 145
  Socrates's, 55, 63
  and tactics of resistance, 145, 150
  and touch, 141–42
  traditions of, 138
  and vision of the whole, 145

conversion moments, ecological, 214–15
Cook, Méira, 259
Corner Brook, Newfoundland
  overview of, 255
  Carl Leggo in, 255–56, 261–62, 266, 269, 274
  Lynch's lane, 262, 269–70
  Margaret McKeon in, 255, 260
  paper mill, 277–78
  winters in, 255–57, 262, 267–69
corporations, 129–31
courage
  climate change, confronting, 58–59
  examples of, 246
  and humility, 59
  Socrates's, 54
  and the state, 60
creation narratives, 5–6, 21
Crozier, Lorna, 266
Culhane, Claire, 245–46
cultivation
  of blue camas, 17–18, *19*
  of clams, 11–12, *13*, 18, 20

Daoism
  Celestial Masters
    cosmic family concept, 103
    history of, 97
    moderation emphasis, 107
    nature, empathy for, 103–7, 111
    and *qi*, 103
  the *Chongyang lijiao shiwu lun*, 108
  climate crisis, lessons for, 111–12
  the *Daode jing*, 99, 101, 110
  *Daojia versus Daojiao*, 97
  diversity of, 97–98
  Dong Zhongshu, 101–3, 114n31
  emotion in, 110
  etymology of, 97
  exploitation, views of, 108

CPSIA information can be obtained
at www.ICGtesting.com
Printed in the USA
LVHW090526180121
676779LV00029B/512